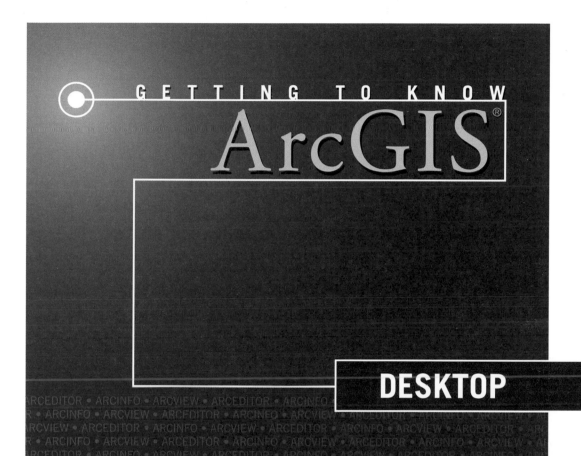

GETTING TO KNOW
ArcGIS®

DESKTOP

Ormsby • Napoleon • Burke • Groessl • Bowden

ESRI PRESS
REDLANDS, CALIFORNIA

Esri Press, 380 New York Street, Redlands, California 92373-8100

Copyright © 2010 Esri

All rights reserved. First edition 2001. Second edition 2004

16 15 14 13 12   4 5 6 7 8 9 10

Printed in the United States of America

Library of Congress Cataloging-in-Publication Data
Getting to know ArcGIS desktop / Timothy James Ormsby...[et al.].—2nd ed.
    p.   cm.
   Includes index.
   ISBN 978-1-58948-260-9 (pbk. : alk. paper)
   1.   ArcGIS. 2.   Geographic information systems. 3.   Graphical user interfaces (Computer systems) I. Ormsby, Tim.
   G70.212.G489 2010
   910.285'53—dc22   2010014231

Ask for Esri Press titles at your local bookstore or order by calling 800-447-9778, or shop online at www.esri.com/esripress. Outside the United States, contact your local Esri distributor or shop online at www.eurospanbookstore.com/esri.

Esri Press titles are distributed to the trade by the following:
In North America:
Ingram Publisher Services
Toll-free telephone: 800-648-3104
Toll-free fax: 800-838-1149

E-mail: customerservice@ingrampublisherservices.com
In the United Kingdom, Europe, Middle East and Africa, Asia, and Australia:
Eurospan Group
3 Henrietta Street
London WC2E 8LU
United Kingdom
Telephone: 44(0) 1767 604972
Fax: 44(0) 1767 601640
E-mail: eurospan@turpin-distribution.com

CONTENTS

ACKNOWLEDGMENTS

ESRI thanks the following people and groups for contributing data and images to this book.

Maps of rainfall, elevation, soils, and growing seasons for North and South America (chapter 1) are based on data provided by DATA+ and the Russian Academy of Sciences.

Maps of elevation, hillshade, and temperature for the Mojave Desert (chapter 5) are based on data provided by the U.S. Defense Department Mojave Desert Ecosystem Program.

Satellite image and map of Mission Bay, San Diego, California, (chapter 5) provided by the U.S. Geological Survey.

African diamond mine, river, and wildlife layers (chapters 5 and 6) provided by DATA+ and the Russian Academy of Sciences.

State of Louisiana layers (chapter 9), including nonhazardous waste pits, parish boundaries, and navigable waterways, provided by David Gisclair of the Louisiana Oil Spill Coordinator's Office, Baton Rouge, Louisiana.

City of Riverside, California, layers (chapter 10), including neighborhoods, places, buildings, and zoning, provided by the City of Riverside GIS Department. City of Riverside census tract layer provided by the U.S. Census Bureau. City of Riverside freeways layer provided by Geographic Data Technology Inc. Copyright 1990–98 Geographic Data Technology Inc. All rights reserved.

Tongass National Forest, Alaska, stream and forest stand layers (chapters 11, 12, and 20) provided by the U.S. Forest Service, Tongass National Forest, Ketchikan Area.

Population attributes for U.S. states and cities (chapter 13) provided by the United States Census Bureau.

City of Manhattan, Kansas, and Riley County, Kansas, layers (chapters 14, 15, and 16), including parcels, water lines, water valves, fire hydrants, and air photo, provided by Dan Oldehoeft, City of Manhattan, Kansas, and John Cowan, Riley County, Kansas.

Satellite image of Crater Lake (chapter 15) provided by the United States Geological Survey. On the Web: `craterlake.wr.usgs.gov/space.html`.

Atlanta streets layer (chapter 17) provided by Geographic Data Technology, Inc. Copyright 1990–98 Geographic Data Technology Inc. All rights reserved.

Typhoon Etang latitude and longitude coordinates (chapter 18) provided by the U.S. Defense Department Joint Typhoon Warning Center.

Tiger reserve layers for India (chapter 18) provided by the World Wildlife Fund. On the Web: `www.worldwildlife.org`.

Thanks to our editor, Michael Karman, *il miglior fabbro*.

Thanks also to Jonell Alvi for additional writing; to Judy Boyd, Tom Brenneman, Nick Frunzi, Christian Harder, Makram Murad, Brian Parr, Gillian Silvertand, Damian Spangrud, Thad Tilton, and Randy Worch for technical reviews and advice; to Prashant Hedao and Brian Parr for data acquisition; to Donna Celso for redesign; to Riley Peake for technical review; and to Michael Law for the ArcGIS 10 update.

INTRODUCTION

Getting to Know ArcGIS Desktop is a workbook for beginners. Its detailed, step-by-step exercises teach you the core functionality of ArcGIS Desktop software: how to make maps, carry out spatial analysis, and build and edit spatial databases in the context of realistic projects. The exercises are supported by conceptual discussions at the start of each chapter and as needed throughout the book. Abundant color graphics confirm your progress along the way.

The exercises can all be completed with an ArcView, ArcEditor, or ArcInfo license. That's because all three software products share a common interface and much of the same functionality: anything that can be done in ArcView can also be done in ArcEditor or ArcInfo, and in exactly the same way. For this reason, *Getting to Know ArcGIS Desktop* is an introduction to each of the products. (Advanced capabilities of ArcEditor and ArcInfo are not covered.)

Getting to Know ArcGIS Desktop has been a GIS best seller since its publication in 2001. This volume has been updated to ensure compatibility with ArcGIS 10, with many new graphics and instructions added to showcase the latest functionality.

The book comes with a CD containing exercise data. To download a trial copy of ArcGIS Desktop, ArcEditor license, go to www.esri.com/esripresss. You need to install both the software and the data to do the exercises in the book. (If you have access to a computer on which the ArcView, ArcEditor, or ArcInfo license of ArcGIS Desktop 10 is already installed, you only need to install the data CD.) Appendix B describes the installation process.

Getting to Know ArcGIS Desktop is a hands-on workbook meant to be a practical manual for classroom lab work or on-the-job training. If you have no GIS background, chapter 1 gives you a quick overview. If you have no ArcGIS software experience, chapter 2 describes ESRI GIS software products.

The book has two introductory chapters and eighteen exercise chapters. Each exercise chapter contains two to four exercises that focus on a particular GIS task or problem. Many common tasks are covered, including symbolizing and labeling maps, classifying data, querying maps, analyzing spatial relationships, setting map projections, building spatial databases, editing data, geocoding addresses, and making map layouts.

Each new exercise in *Getting to Know ArcGIS Desktop* is a fresh starting point, with the maps and data you need already prepared for you. It is advisable to follow the chapters in order because tools and functions used often in early chapters may not be described again in later ones. The exercises will work, however, no matter which chapter you start with. Each chapter takes about an hour or two to complete.

Roll up your sleeves and start *Getting to Know ArcGIS Desktop*.

Chapter 1

Introducing GIS

For a long time, people have studied the world using models such as maps and globes. In the past thirty years or so, it has become possible to put these models inside computers—more sophisticated models into smaller computers every year. These computer models, along with the tools for analyzing them, make up a geographic information system (GIS).

In a GIS, you can study not just this map or that map, but every possible map. With the right data, you can see whatever you want—land, elevation, climate zones, forests, political boundaries, population density, per capita income, land use, energy consumption, mineral resources, and a thousand other things—in whatever part of the world interests you.

The map of the world, below, shows countries, cities, rivers, lakes, and the ocean.

The map has a legend (or table of contents) on the left and a display area on the right.

A GIS map contains layers

On a paper map, you can't peel cities away from countries, or countries away from the ocean, but on a GIS map you can. A GIS map is made up of layers, or collections of geographic objects that are alike. To make a map, you can add as many layers as you want.

This world map is made up of five layers. It could have many more.

Layers may contain features or surfaces

In the map (page 2), the Cities layer includes many different cities and the Rivers layer many different rivers. The same is true of the Lakes and Countries layers. Each geographic object in a layer—each city, river, lake, or country—is called a feature.

Not all layers contain features. The Oceans layer is not a collection of geographic objects the way the others are. It is a single, continuous expanse that changes from one location to another according to the depth of the water. A geographic expanse of this kind is called a surface.

Features have shape and size

Geographic objects have an endless variety of shapes. All of them, however, can be represented as one of three geometrical forms—a polygon, a line, or a point.

Polygons represent things large enough to have boundaries, such as countries, lakes, and tracts of land. Lines represent things too narrow to be polygons, such as rivers, roads, and pipelines. Points are used for things too small to be polygons, such as cities, schools, and fire hydrants. (The same object may be represented by a polygon in one layer and a line or a point in a different layer, depending on how large it is presented.)

Polygons, lines, and points collectively are called vector data.

Surfaces have numeric values rather than shapes

Unlike countries or rivers, things such as elevation, slope, temperature, rainfall, and wind speed have no distinct shape. What they have instead are measurable values for any particular location on the earth's surface. (Wherever you go, for instance, you are either at sea level or a number of meters above or below it.) Geographic phenomena like these are easier to represent as surfaces than as features.

The most common kind of surface is a raster, a matrix of identically sized square cells. Each cell represents a unit of surface area—for example, 10 square meters—and contains a measured or estimated value for that location.

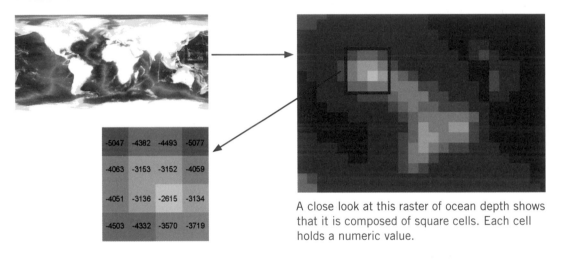

A close look at this raster of ocean depth shows that it is composed of square cells. Each cell holds a numeric value.

The world is not divided neatly into features and surfaces. Many things can be looked at either way. For example, polygons are often used to mark the boundaries of different vegetation types in a region, but this implies that the change from one type to another is more abrupt than it probably is. Vegetation can also be represented as a raster surface, where each cell value stands for the presence of a type of vegetation.

Features have locations

If you were asked to find Helsinki, Finland, on a map of the world, it probably wouldn't take you very long. But suppose Helsinki wasn't shown on the map. Could you make a pencil mark where it ought to go?

Now suppose you could lay a fine grid over the world map and you knew that Helsinki was a certain number of marks up from and to the right of a given starting point. It would be easy to put your pencil on the right spot. A grid of this kind is called a coordinate system, and it's what a GIS uses to put features in their proper places on a map.

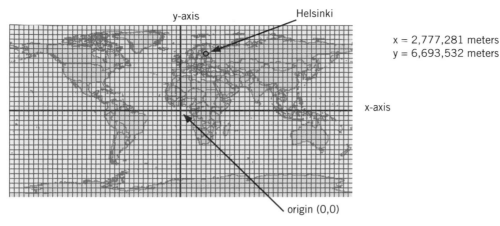

y-axis   Helsinki

x = 2,777,281 meters
y = 6,693,532 meters

x-axis

origin (0,0)

On a map, a coordinate system has an x-axis and a y-axis. The intersection of the axes is called the origin. Feature locations are specified by their distance from the origin in meters, feet, or a similar unit of measure.

The location of a point feature on a map is defined by a pair of x,y coordinates. A straight line needs two pairs of coordinates—one at the beginning and one at the end. If the line bends, like a river, there must be a pair of coordinates at every location where the line changes direction. The same holds true for a polygon, which is simply a line that returns to its starting point.

Features can be displayed at different sizes

On a GIS map, you can zoom in to see features at closer range. As you do so, the scale of the map changes.

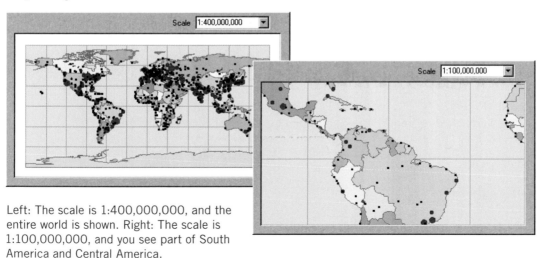

Left: The scale is 1:400,000,000, and the entire world is shown. Right: The scale is 1:100,000,000, and you see part of South America and Central America.

Scale, commonly expressed as a ratio, is the relationship between the size of features on a map and the size of the corresponding places in the world. If the scale of a map is 1:100,000,000, it means that features on the map are one hundred million times smaller than their true size.

Zooming in gives you a closer view of features within a smaller area. The amount of detail in the features does not change, however. A river has the same bends, and a coastline the same crenulations, whether you are zoomed in and can discern them or are zoomed out and cannot.

How much detail features have depends on the layer you use. Just as a paper map of the world generalizes the shape of Brazil more than a map of South America does, so different GIS layers can contain more feature detail or less.

Features are linked to information

There is more to a feature than its shape and location. There is everything else that might happen to be known about it. For a country, this might include its population, capital, system of government, leading imports and exports, average rainfall, mineral resources, and many other things. For a road, it might be its speed limit, the number of lanes it has, whether it is paved or unpaved, and whether it is one-way or two-way. There is a great deal of information to be had about any feature, from a humble length of sewer pipe to an ocean.

Information about the features in a layer is stored in a table. The table has a record (row) for each feature in the layer and a field (column) for each category of information. These categories are called attributes.

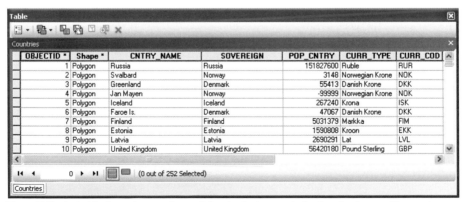

The attribute table for a layer of countries includes each feature's shape, ID number, and name, among other things.

Features on a GIS map are linked to the information in their attribute table. If you highlight China on a map, you can bring up all the information stored about it in the attribute table for countries. If you highlight a record in the table, you see the corresponding feature on the map.

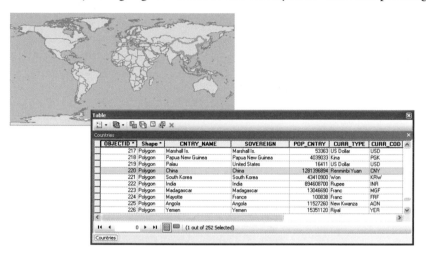

The link between features and their attributes makes it possible to ask questions about the information in an attribute table and display the answer on the map.

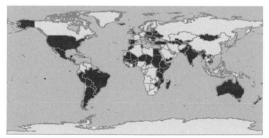
Which cities are national capitals?

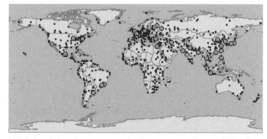
Which cities have populations over five million?

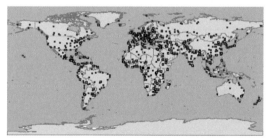
Which countries are net importers of goods?

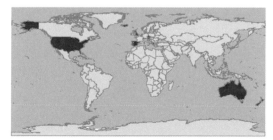
Which countries are net importers of goods and have per capita GDP of $10,000 or more?

Similarly, you can use attributes to create thematic maps, maps in which colors or other symbols are applied to features to indicate their attributes.

Energy consumption per capita

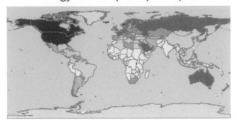

The darker the shade of brown, the more energy is used per person in each country.

Migration

Red countries have net emigration, blue countries have net immigration. Light yellow countries have little or no change.

Urban population by percentage

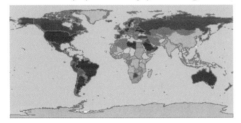

Darker shades of purple show countries where a higher percentage of the population lives in cities.

Greenhouse gas emissions

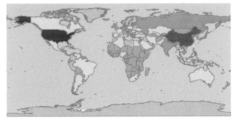

Greenhouse gas emissions are lowest in green countries, higher in yellow and orange countries, and highest in red countries.

Features have spatial relationships

Besides asking questions about the information in attribute tables, you can also ask questions about the spatial relationships among features—for example, which ones are nearest others, which ones cross others, and which ones contain others. The GIS uses the coordinates of features to compare their locations.

Which cities are within 50 kilometers of a river?

Which countries have a river that crosses their border?

Which countries share a border with China?

Which countries contain a lake completely within their borders?

New features can be created from areas of overlap

Questions about attributes and spatial relationships identify existing features that do or do not have certain qualities. To solve some geographic problems, however, a GIS must create new features. Suppose you want to find suitable places for growing amaranth, a nutritious grain originally grown by the Aztecs. You know that in Mexico amaranth is grown in areas of moderate elevation (1,000 to 1,500 meters) that have an average yearly rainfall of 500 to 800 millimeters, and loam or sandy-loam soil. You also know that the plant requires a fairly long growing season, at least 120 days without frost.

You have layers of elevation, rainfall, growing season, and soil type for North and South America.

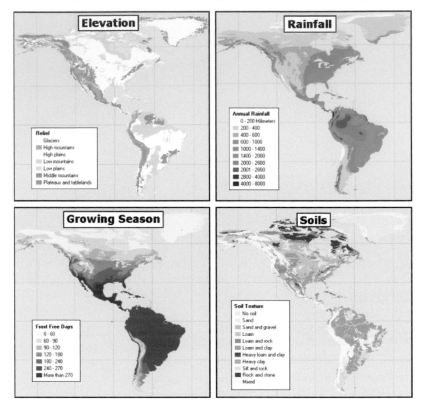

To find places that meet the specified conditions, the GIS looks for areas of overlap among features in the different layers. Wherever there is overlap among four features with the right attributes—the right elevation, the right amount of rainfall, the right growing season, and the right kind of soil—a new feature is created. The new feature's boundary is the area of overlap, which is different from the boundaries of each of the four features it was created from.

The result of the analysis is a new layer that shows where amaranth can be grown.

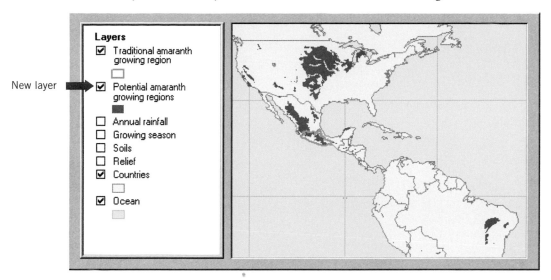

You now have some idea of what a GIS is and what it can do. In the next chapter, you'll learn a little about ArcGIS Desktop, the latest GIS software from ESRI.

Chapter 2

Introducing ArcGIS Desktop

When you shop for a car, a cordless vacuum, or a credit card, you often choose among different models in a product line. The models do essentially the same thing, and usually look similar—the difference is in the extras that come with the medium or high-end models.

ArcGIS Desktop is a GIS software product line. The entry-level model is ArcReader, a product that comes with every ArcGIS installation or can be downloaded for free from the ESRI Web site. ArcReader lets you view and print digital maps created by others, navigate to different parts of the map, and zoom to various levels of detail.

ArcView makes the maps and data that ArcReader can only view and print. With ArcView, you can also query data; analyze spatial relationships like distance, intersection, and containment among map features; and overlay layers to discover how different types of data are interrelated at particular locations.

ArcEditor gives you complete ArcView functionality and has additional data creation and editing tools. Other features include versioning, which allows multiple users to edit data simultaneously, and disconnected editing, which allows data to be checked out of a database, edited in the field, and then checked back in.

At the top of the line, ArcInfo gives you complete ArcEditor functionality plus a full set of spatial analysis tools. ArcInfo also comes bundled with ArcInfo Workstation, a parallel, self-standing GIS software product.

ArcView, ArcEditor, and ArcInfo share a common user interface and can freely exchange maps and data.

ArcMap has a data view for creating, symbolizing, and analyzing maps.

ArcMap also has a layout view for composing maps for printing. You can add titles, scale bars, legends, and other elements.

The ArcMap and ArcCatalog applications

GIS tasks can be broadly divided into two categories. One includes mapmaking, editing, and spatial analysis; the other includes database design and data management. This division is reflected in the ArcMap and ArcCatalog applications of which ArcView, ArcEditor, and ArcInfo are composed.

ArcMap is the application for making maps and analyzing data.

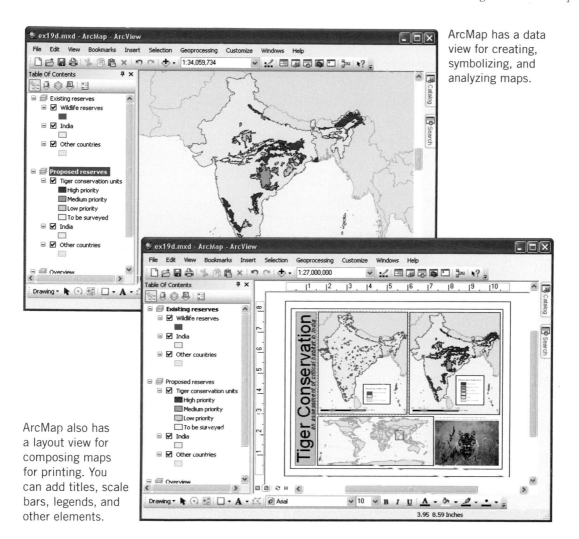

ArcMap has a data view for creating, symbolizing, and analyzing maps.

ArcMap also has a layout view for composing maps for printing. You can add titles, scale bars, legends, and other elements.

ArcCatalog is the data-management application.

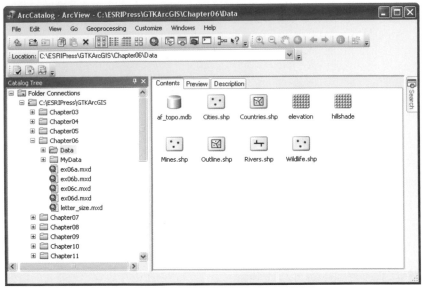

With ArcCatalog, you organize, browse, document, and search for spatial data. ArcCatalog also has tools for building and managing GIS databases.

Extending ArcGIS Desktop

You can augment the capabilities of ArcGIS Desktop with specialized extension products that are fully integrated with ArcView, ArcEditor, and ArcInfo. ArcGIS Spatial Analyst, ArcGIS 3D Analyst, and ArcGIS Geostatistical Analyst, described below, along with several other extension products are included on the demo CD that comes with this book. The book does not teach you how to use them, but you can explore them on your own using the online help and the ArcGIS Tutorial data that is also on the CD. Like ArcView, the extensions are good for 180 days.

ArcGIS Spatial Analyst maps and analyzes measured data like elevation, rainfall, or chemical concentrations. By dividing geographic space into a matrix of square cells that store numbers, ArcGIS Spatial Analyst allows you to represent, query, and statistically summarize this kind of data. You can also estimate values at unmeasured locations through the mathematical interpolation of known sample values.

ArcGIS 3D Analyst gives you the ability to see spatial data in three dimensions. You can "fly through" terrain and examine it from any height or angle. You can model cities and neighborhoods by drawing buildings at their correct heights. Analysis tools let you solve visibility, volume, and downhill path problems. (Which areas can be seen from the mountaintop? Can Tower 1 be seen from Tower 2? How much earth is in the hill? Which course will water follow down a surface?)

ArcGIS Geostatistical Analyst lets you evaluate measured spatial data according to statistical principles. You can explore the value distributions of datasets, compare them to normal (bell-shaped) distributions and to each other, and look for correlations between different types of data. As with ArcGIS Spatial Analyst, you can make maps of predicted values at unmeasured locations. ArcGIS Geostatistical Analyst, however, gives you a wider choice of predictive models, more control over their parameters, and statistical techniques for assessing the quality of the results.

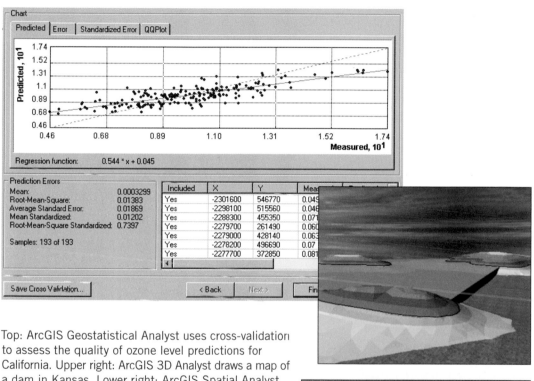

Top: ArcGIS Geostatistical Analyst uses cross-validation to assess the quality of ozone level predictions for California. Upper right: ArcGIS 3D Analyst draws a map of a dam in Kansas. Lower right: ArcGIS Spatial Analyst represents distances to the nearest airport (green points) with graduated colors. (Only airports serving more than 25,000 passengers annually are shown.)

ArcGIS Desktop is just one of five parts of the ArcGIS system, which also includes mobile GIS, online GIS, server GIS, and ESRI Data.

Mobile GIS goes into the field

Mobile GIS technology extends GIS beyond the office and allows organizations to make accurate, real-time business decisions and collaborate in both field and office environments. Mobile GIS products and services are essential to the success of many GIS field mapping applications and enable organizations to expand their enterprise GIS to various mobile platforms. Many industries and field-workers (surveyors, field technicians, delivery staff, public service and utility workers) use mobile GIS technology to help complete both complex projects and routine field tasks. ESRI's mobile GIS products include ArcPad and ArcGIS Mobile for the Windows XP, Vista, 7 and Windows Mobile platforms, and ArcGIS for iPhone which can be accessed and used on the Apple iPhone and iPad platform.

Online GIS goes to digital worlds

ArcGIS Online content and capabilities are built-in to the ArcGIS user experience. ArcGIS Online, hosted by ESRI and powered by ArcGIS Server and ArcGIS Data Appliance, provides ready-to-use, high-quality basemaps, layers, tools, and other content that has been published by ESRI and the user community.

Server GIS goes over the Web and through the enterprise

ArcGIS Server is used to publish the maps, tools, and GIS data created in ArcGIS Desktop as services that can be delivered over the Web and throughout the enterprise. These GIS services can be used in desktop, mobile and Web applications.

ArcGIS Server also manages the exchange of information between ArcGIS and relational database management systems (RDBMs), such as Microsoft SQL Server, IBM DB2, Oracle and PostgreSQL. It enables users to share and edit centrally-stored geographic data over the Web.

ESRI Data goes to work for you

ESRI Data encompasses updated demographics, business, and consumer spending data used by organizations to analyze markets, profile customers, evaluate competitors, and more.

ESRI StreetMap Premium is an enhanced street dataset that provides routing, geocoding, and high-quality cartographic display for the United States, Canada, and Europe.

ESRI Data & Maps is a set of annual map data that is included at no additional cost with ArcGIS software and preconfigured to work with ArcGIS products.

In this book, you will learn the basics of ArcView, ArcEditor, and ArcInfo. To find out more about ArcGIS Desktop extensions, or about mobile GIS, online GIS, server GIS, or ESRI Data, visit the ESRI Web site at **www.esri.com**.

Chapter 3

Exploring ArcMap

Displaying map data
Navigating a map
Looking at feature attributes

ArcMap is an application for displaying maps and investigating them, for analyzing maps to answer geographic questions and producing maps that make analysis persuasive. The ArcMap application window consists of a map display for viewing spatial data, a table of contents for listing the layers shown in the display, and a variety of toolbars for working with the data.

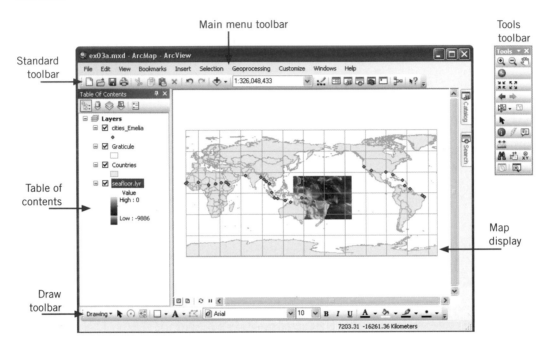

You can change the way ArcMap looks to suit your preferences and the kind of work you do. Toolbars can be hidden or shown. New commands can be added to them. They can be docked at different places in the application window or can float independently of it.

To dock a floating toolbar, drag it to the interface. To undock it, click the vertical gray bar at its left edge and drag it away from the interface. To hide or show a toolbar, click the Customize menu, point to Toolbars, and check or uncheck the toolbar name.

Alternatively, some windows (such as Catalog or Search) can be docked to the interface. These windows have the ability to collapse into a tab or expand to show its entirety. To dock a floating window, drag it to the interface. The window turns blue and four arrows point at locations where the window can be docked. When the window is docked, click the Auto Hide button to enable the tabbed behavior. To remove the tab, click the Auto Hide button and undock the window.

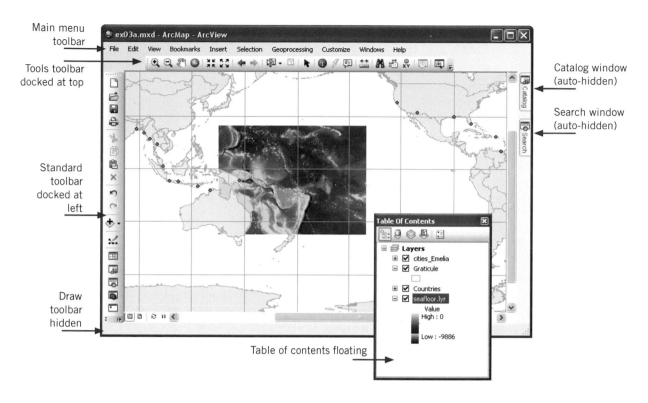

Main menu toolbar

Tools toolbar docked at top

Standard toolbar docked at left

Draw toolbar hidden

Catalog window (auto-hidden)

Search window (auto-hidden)

Table of contents floating

This book assumes that you are working with the default interface. The exercise graphics reflect this, with one exception—toolbars are always shown horizontally.

Changes you make to the interface are applied to subsequent ArcMap sessions, so if you dock a toolbar or window in one session, it will be docked the next time you start ArcMap, and if you resize the application window, it will keep the new size in the next session. Changes like this will not significantly affect the exercises, but may give you slightly different results for such operations as labeling that are influenced by the size and scale of the map display.

For more information about customizing the interface, click the Contents tab in ArcGIS Desktop Help and navigate to *Customizing and developing with ArcGIS > Customizing the user interface.*

Displaying map data

In this exercise, you'll learn how to display data in ArcMap. You'll learn how to navigate maps and get information about map features.

You will be able to do the exercises only if you have installed the 180-day trial version of ArcView 10 that comes with this book, or if you have a licensed version of ArcEditor 10 or ArcInfo 10 software on your computer. Keep in mind that using a previous version of ArcGIS means that certain tools, functions, windows, or dialog boxes may or may not be present. You may encounter steps in the exercises where a workaround may be necessary.

Exercise 3a

You work for an aviation history foundation that is researching the last flight of Amelia Earhart. In 1937, Earhart was near the end of a flight around the world when her plane disappeared over the Pacific Ocean. The U.S. government spent more than $4 million searching for Earhart and her navigator, Fred Noonan. The foundation believes Earhart may have crashed on Nikumaroro, one of several tiny islands that make up the country of Kiribati, and would like to mount an expedition to look for the wreckage. You have been asked to manage a GIS project that will help organize data and acquaint potential sponsors with the foundation's plans.

1 Start ArcMap by clicking the Start button on the Windows taskbar, point to All Programs, point to ArcGIS 10, and click ArcMap 10.

When ArcMap opens, you see the Getting Started dialog box on top of the main application window. This dialog box allows you to quickly start a new map, open an existing map or template.

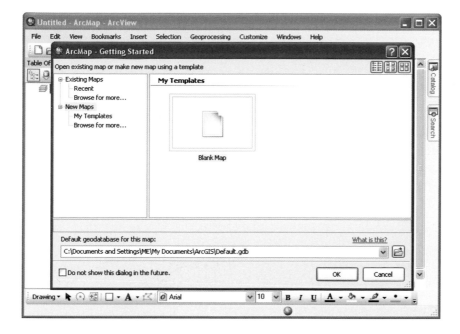

2 In the Getting Started dialog box, under the Existing Maps section, click "Browse for more…"

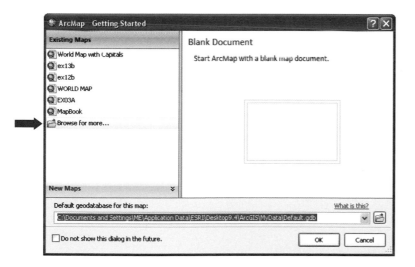

3   In the Open dialog box, navigate to **C:\ESRIPress\GTKArcGIS\Chapter03** (or to the folder where you installed the GTKArcGIS data). Click **ex03a.mxd**, as shown in the following graphic, and click Open.

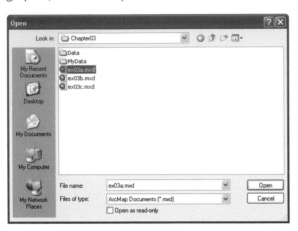

A preview of the map document opens inside the New Document dialog box.

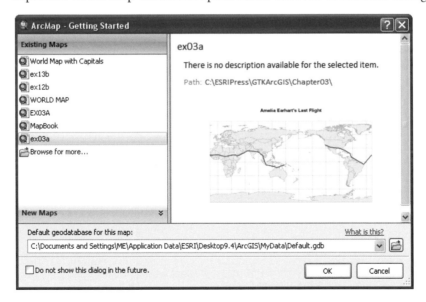

4   In the New Document dialog box, click OK.

The map document opens. The map looks different from many world maps because it is centered on the South Pacific area where Amelia Earhart vanished, rather than on the prime meridian (which runs through Greenwich, England).

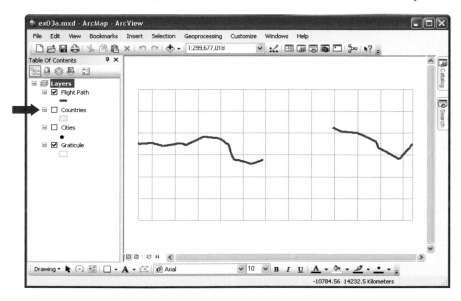

The map shows the countries of the world, Earhart's flight path, and a graticule, which is the technical name for lines of latitude and longitude on a map. Each of these categories of geographic information (countries, flight path, graticule) is called a layer.

The table of contents lists the names of the layers in the map. It shows the color or symbol used to draw each layer and tells you, by a check mark, whether or not the layer is visible. The Flight Path, Countries, and Graticule layers are currently visible. The Cities layer is not.

5   In the table of contents, click the check box next to the Countries layer to turn it off.

The countries disappear from the map.

6   In the table of contents, click the Countries check box to turn it back on. Click the check box next to the Cities layer to turn it on as well.

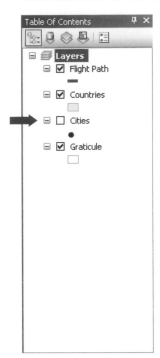

The Cities layer is checked but you still can't see the cities. This is because data is displayed on the map in the order of the layers in the table of contents. The Cities layer is covered by the countries.

7   In the table of contents, click the Cities layer name to highlight it. Click and drag the layer to the top of the table of contents, then release the mouse button. As you drag the layer, a horizontal black bar indicates its position.

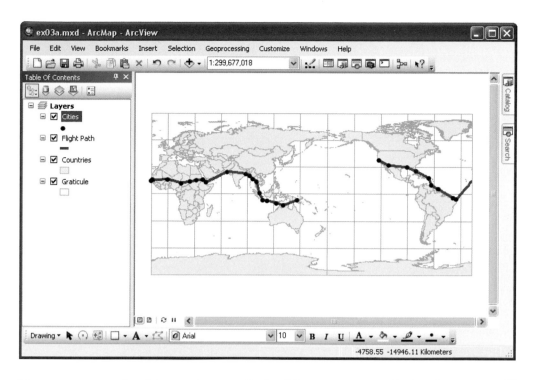

Now you can see the cities on the map. Each of them is a stop on Earhart's route. You'll change the layer name to make this clear.

8   In the table of contents, right-click the Cities layer name. A context menu opens. Many ArcMap operations are started from context menus. On the context menu, click Properties to open the Layer Properties dialog box.

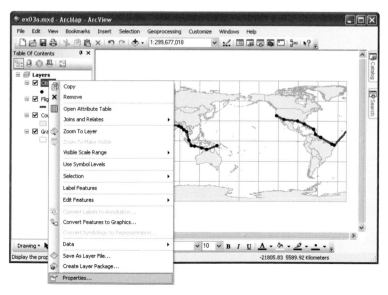

9 In the Layer Properties dialog box, click the General tab.

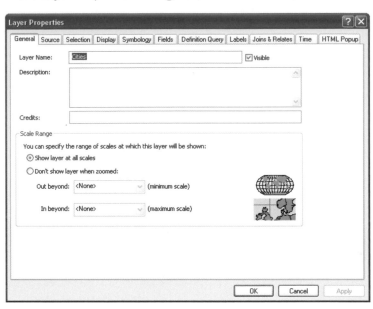

The Layer Properties dialog box has several tabs for setting layer properties, most of which you'll use in this book.

10 In the Layer Name text box, the name "Cities" is highlighted. Type **Cities Earhart Visited** in its place. Make sure that your dialog box matches the following graphic, then click OK.

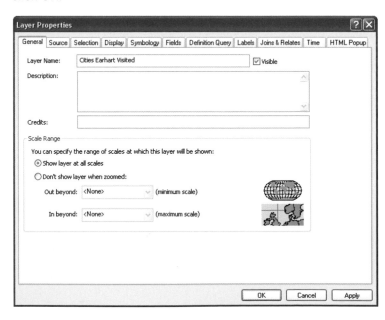

The layer is renamed in the table of contents.

You will now track the progress of Earhart's round-the-world flight, which began on the west coast of the United States. She chose an equatorial route so as to circle the globe at its full circumference and flew east to minimize the effects of storms and headwinds.

11 On the Tools toolbar, click the Zoom In tool. (Move the mouse pointer over it to see its name.) Your toolbar may be oriented vertically. You can change its orientation if you like by dragging one of its corners.

12 Move the mouse pointer over the map. The cursor changes to a magnifying glass. Drag a box around the United States, approximately as shown in the following graphic. (If you make a mistake, click either the Full Extent or Go Back To Previous Extent buttons and try again.)

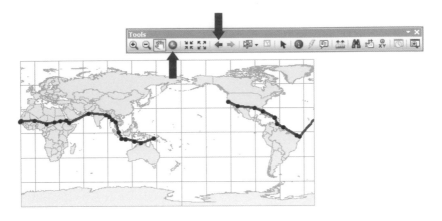

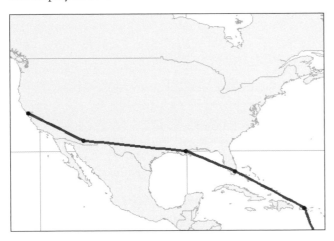

The display zooms in on the United States.

Zooming in or out changes the display scale, which is shown on the Standard toolbar. When the map showed the whole world, the scale was about 1:300,000,000. This means that map features are displayed at one three hundred millionths of their actual size. The scale should now be about 1:50,000,000. (Scale is also affected by the size of the ArcMap application window.)

Although the cities are not labeled, you can find out their names and get other information about them with ArcMap tools.

13 On the Tools toolbar, click the Select Elements tool.

14 Move the cursor over the westernmost city on the display. The city name displays as a map tip.

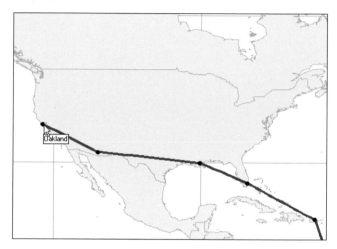

Map tips are a layer property. They can be turned on or off on the Display tab of the Layer Properties dialog box. You can see map tips no matter which tool is selected.

15 Move the cursor over the other three cities on the display. They are Tucson, New Orleans, and Miami. You may also be able to see San Juan, Puerto Rico.

To make the names visible at all times, you can label the cities.

16 In the table of contents, right-click the Cities Earhart Visited layer and click Label Features.

The name of each city appears next to the map feature. From Miami, Earhart flew southeast to Puerto Rico and then to South America.

17 On the Tools toolbar, click the Pan tool.

18 Move the mouse pointer over the map. The cursor changes to a hand. Click and drag the display up and to the left until Miami is in the upper left corner of the window. Release the mouse button.

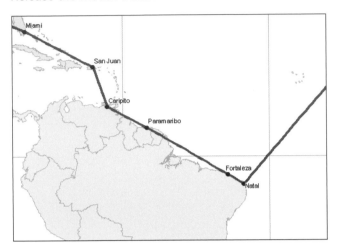

You can get information about any of the cities Earhart visited with the Identify tool.

19 On the Tools toolbar, click the Identify tool. When you click the tool, the Identify window opens. If it covers most of the map display, move it out of the way.

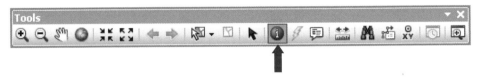

20 On the map, click the city of Natal. You must click exactly on the city or you will identify something else, such as a segment of the flight path or the country of Brazil. If this happens, try again.

The Identify window shows you the country and city name, and various facts about Natal, such as its elevation (meters), its average annual precipitation (millimeters), and the dates Earhart and Noonan arrived and departed.

21 Close the Identify window. On the Tools toolbar, click the Full Extent button.

The map zooms to its original extent.

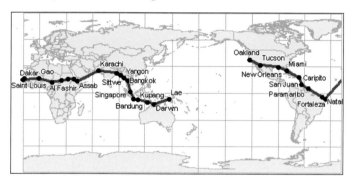

In the next exercise, you will follow the rest of Earhart's journey to the point where she and Noonan disappeared.

22 If you want to save your work, click the File menu and click Save As. Navigate to **\GTKArcGIS\Chapter03\MyData**. Rename the file **my_ex03a.mxd** and click Save.

Saving your work is optional. When you begin a new exercise, however, it is important that you open the new map document specified in the instructions. The exercises are not always perfectly continuous, and if you work in the same map document throughout a chapter you may get results that don't match those in the book.

23 If you are continuing with the next exercise, leave ArcMap open. Otherwise, click the File menu and click Exit. Click No if prompted to save your changes.

Navigating a map

In this exercise, you'll continue to work with ArcMap navigational tools. You will also learn how to create spatial bookmarks, which save a specific view of a map.

Exercise 3b

Earhart and Noonan crossed the Atlantic at night. When they saw the west coast of Africa, they realized they were north of their intended destination—the city of Dakar in Senegal. They landed at the first airstrip they saw, in the Senegalese city of Saint Louis, and from there made the short flight to Dakar. They proceeded to fly across Africa and Asia, making their last stop in Lae, Papua New Guinea. They intended to go on to tiny Howland Island in the South Pacific, then to Hawaii and back to California.

1   Start ArcMap. In the ArcMap—Getting Started dialog box, under the Existing Maps section, click Browse for more. (If ArcMap is already running, click the File menu and click Open.) Navigate to **C:\ESRIPress\GTKArcGIS\Chapter03**. Click **ex03b.mxd** and click Open.

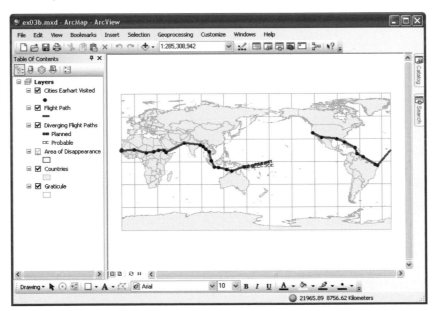

The map resembles the one in the previous exercise but has two additional layers. The Diverging Flight Paths layer contains two line features. One represents the course Earhart and Noonan planned to take. The other represents the course the foundation believes they actually followed.

The Area of Disappearance layer is shown in the table of contents with a grayed-out check mark (and a tiny scale bar under the check box). This means that the layer's visibility depends on the map's display scale. This layer includes hundreds of Pacific islands too small to be represented on a general world map. When you zoom in to look at the end of Earhart's flight, this layer will become visible.

2   On the Tools toolbar, click the Zoom In tool.

3   On the map, drag a rectangle that includes the northern half of Africa and reaches to India, as shown in the following graphic.

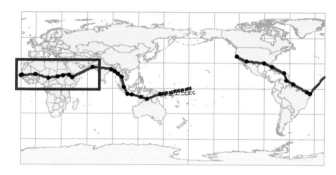

The display zooms in on the African stretch of the route. If your display doesn't show the west coast of Africa, use the Pan tool to adjust it.

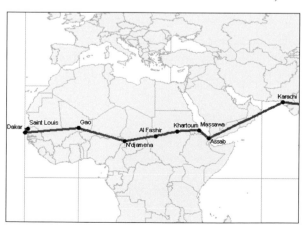

At this scale, the city labels appear. The display of labels, like that of map features, can be made scale-dependent in the Layer Properties dialog box. In this map, labels appear only when the scale is larger than 1:100,000,000. (The larger the scale, the nearer features are to their actual size.)

4   On the Tools toolbar, click the Pan tool.

5   Drag the display to the left to follow the flight path.

From Assab (in what is now Eritrea) on the east coast of Africa, Earhart flew to Karachi (now in Pakistan). She then headed south, flying over Southeast Asia to Indonesia.

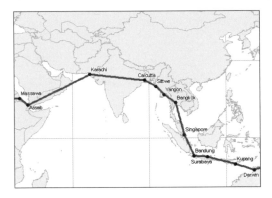

6   Continue panning along the route.

From Darwin, Australia, Earhart flew to her last known stop in Lae, Papua New Guinea. At this point, the path the foundation believes she followed diverges from the planned flight path.

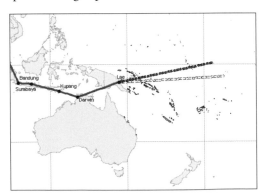

In the table of contents, the check mark by the Area of Disappearance layer is no longer grayed-out. (The layer displays at scales larger than 1:100,000,000.) The features in this layer are small islands. Until you zoom in very close, you see mostly outlines.

7 In the table of contents, right-click the Area of Disappearance layer and click Zoom To Layer.

You can see a number of tiny islands in the display, but it looks as if both flight paths stop in the middle of the ocean.

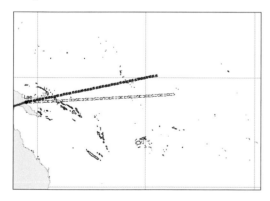

8 Click the Window menu and click Magnifier. A magnification window opens on top of the display. (Yours may open in a slightly different position from the one in the graphic.)

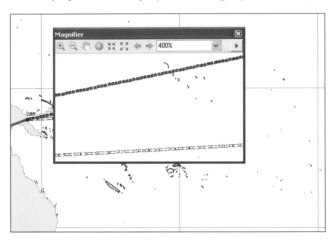

When you move the Magnifier window over the map, the area seen through the window is magnified four times (or 400 percent).

9 Click and drag the blue title bar at the top of the Magnifier window to move it to the right. As you drag it, it displays a crosshair to show you the point on which it's centered. Place the crosshair on the end of the probable flight path and release the mouse.

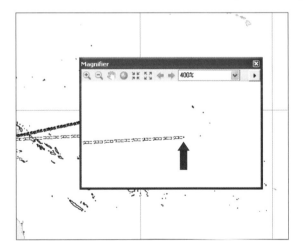

If your eyes are good, you can discern the tiny island of Nikumaroro at the end of the probable flight path. If not, you will soon get a better look.

10 Close the Magnifier window. On the Tools toolbar, click the Zoom In tool.

11 On the map, drag a rectangle that includes the ends of both the planned and probable flight paths.

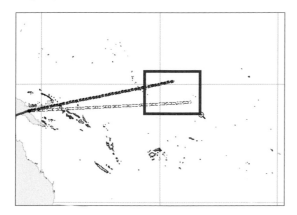

The display zooms in.

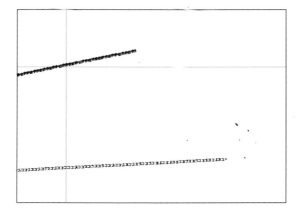

Even at this scale, it's hard to see land. Earhart and Noonan disappeared somewhere in this watery expanse. You will set a bookmark to save this map extent. You'll return to it later to measure the distance from Howland Island to Nikumaroro—the distance, if the foundation is right, by which the aviators were off course.

12 Click the Bookmarks menu and click Create.

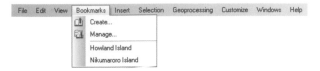

The Spatial Bookmark dialog box opens.

13 Replace the existing text with **End of Flight**, as shown in the following graphic, then click OK.

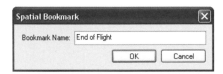

Now you'll zoom in for a close look at the islands at the ends of the two flight paths. You'll use bookmarks that have already been created.

14 Click the Bookmarks menu and click Howland Island.

The display zooms in on Howland Island at the end of the planned flight path. Earhart intended to refuel here before going on to Hawaii.

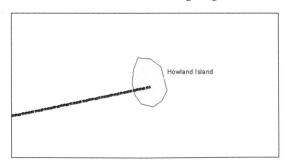

15 Click the Bookmarks menu and click Nikumaroro Island. The display zooms in on Nikumaroro Island.

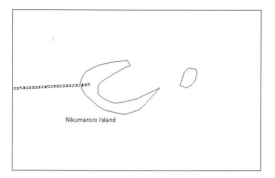

Clothing and empty food tins found on the west bank of the island suggest that Earhart and Noonan may have crashed nearby. In addition, Earhart said, in her last message to the U.S. Coast Guard cutter Ithaca, "We are in line of position 158 degrees –337 degrees…" The line she mentioned is a sun line, used in celestial navigation, that runs directly through both Howland Island and Nikumaroro.

If, indeed, Earhart crashed on or near Nikumaroro, she and Noonan were far off course.

16 Click the Bookmarks menu and click End of Flight. The display zooms to the extent you bookmarked.

17 On the Tools toolbar, click the Measure tool. The Measure window opens. Click the Choose Units drop-down arrow, point to Distance, and click Kilometers.

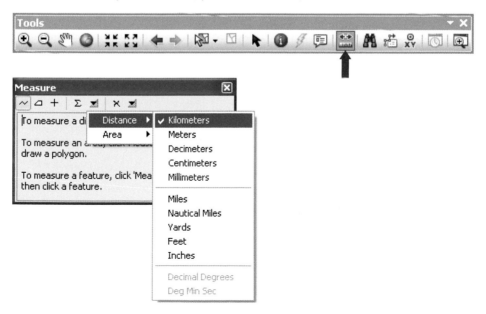

18 Move the mouse pointer over the display. The cursor is a ruler with a small crosshair. Place the crosshair at the end of the planned flight path and click to begin a line. Move the cursor to the end of the probable flight path and double-click to end the line.

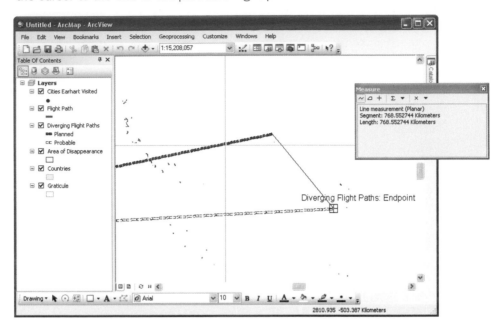

The length of the line is displayed in the Measure window. If the probable path is the actual one, Earhart and Noonan were off course by about 770 kilometers. Poor maps, cloud cover, the scarcity of landmarks, and the sheer length of the planned flight from Lae all may have contributed.

19 If you want to save your work, click the File menu and click Save As. Navigate to **\GTKArcGIS\Chapter03\MyData**. Rename the file **my_ex03b.mxd** and click Save.

20 If you are continuing with the next exercise, leave ArcMap open. Otherwise, click the File menu and click Exit. Click No if prompted to save your changes.

Looking at feature attributes

In a GIS, a feature on a map may be associated with a great deal of information—more than can be displayed at any given time. This information is stored in an attribute table. A layer's attribute table contains a row (or record) for every feature in the layer and a column (or field) for every attribute or category of information.

When you clicked the city of Natal to identify it at the end of exercise 3a, the information you saw in the Identify window was the information stored in the layer attribute table.

In this exercise, you will look at the attribute tables for two map layers. You will learn how to change a table's appearance and how to get statistical information from it.

Exercise 3c

The long transatlantic flight from Brazil to Senegal put Earhart north of her intended destination. The flight from Lae to Howland Island would have been even longer. You will look at the lengths of the various stages of the flight.

1   Start ArcMap. In the ArcMap—Getting Started dialog box, under the Existing Maps section, click Browse for more. (If ArcMap is already running, click the File menu and click Open.) Navigate to **C:\ESRIPress\GTKArcGIS\Chapter03**. Click **ex03c.mxd** and click Open.

You see the familiar map of the world and of Earhart's flight.

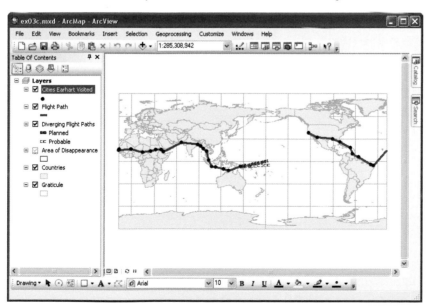

Now you will open the attribute table for the Cities Earhart Visited layer.

2   In the table of contents, right-click the Cities Earhart Visited layer and click Open Attribute Table.

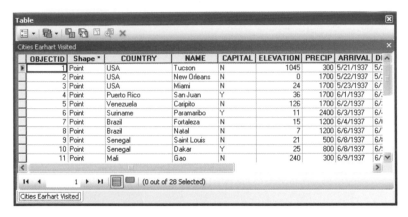

3   Scroll down through the table. There are twenty-eight records, one for each city. Scroll across the table to look at the attributes.

There are ten attributes, or fields. The OBJECTID field contains a unique identification number for every record. The Shape field describes the feature geometry. Among the other attributes are the name of each city, the dates Earhart arrived and departed, and comments on any unusual activity.

The intersection of a record and a field is a cell. A cell contains an attribute value. For example, the attribute value of the NAME field for the first record is "Tucson."

You'll adjust the display width of the fields so you can see more attributes.

4   Scroll back all the way to the left. Place the mouse pointer on the vertical black bar between the NAME and CAPITAL fields. The cursor changes to a two-headed arrow.

5 Drag the cursor to the left. As you drag, the original field width is marked by a vertical red line and the new width by a vertical black line. Release the mouse button somewhere before you start cutting off city names.

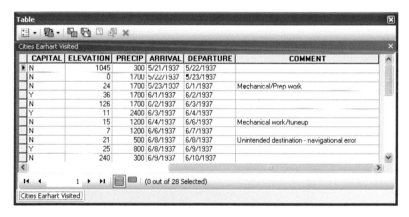

The last letter of the DEPARTURE field name is cut off. You'll widen this field.

6 Scroll to the right side of the table. Place the mouse pointer on the vertical black bar between the DEPARTURE and COMMENT fields. Drag the cursor slightly to the right to see the entire field name.

CAPITAL	ELEVATION	PRECIP	ARRIVAL	DEPARTURE	COMMENT
N	1045	300	5/21/1937	5/22/1937	
N	0	1700	5/22/1937	5/23/1937	
N	24	1700	5/23/1937	6/1/1937	Mechanical/Prep work
Y	36	1700	6/1/1937	6/2/1937	
N	126	1700	6/2/1937	6/3/1937	
Y	11	2400	6/3/1937	6/4/1937	
N	15	1200	6/4/1937	6/6/1937	Mechanical work/tuneup
N	7	1200	6/6/1937	6/7/1937	
N	21	500	6/8/1937	6/8/1937	Unintended destination - navigational error
Y	25	800	6/8/1937	6/9/1937	
N	240	300	6/9/1937	6/10/1937	

(0 out of 28 Selected)

Cities Earhart Visited

The elevation and precipitation (ELEVATION and PRECIP) fields could contribute to a study of the weather Earhart faced. The CAPITAL field is probably not useful for any analysis connected with the flight. You will hide this field.

7   Place the mouse pointer on the vertical black bar between the CAPITAL and ELEVATION fields. Click and drag the cursor to the right edge of the NAME field and release the mouse button. The CAPITAL field is hidden.

If you wanted to restore the hidden field, you would double-click the border between the NAME and ELEVATION field names.

You can rearrange the order of fields as well. It would be more natural to have the city name appear before the country name.

8   Scroll to the left. Click the column heading of the NAME field. The field is highlighted.

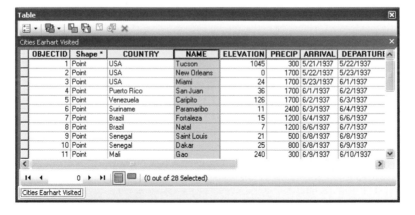

9   Drag the column heading to the left. The cursor becomes a pointer with a small rectangle, showing that a field is being moved. When the vertical red line is between the Shape and COUNTRY fields, as shown in the following graphic, release the mouse button.

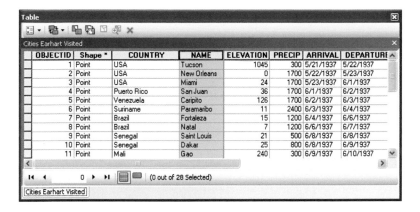

The field names are rearranged.

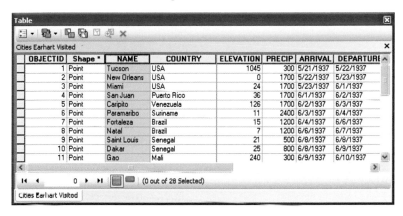

Records, as well as fields, can be highlighted. When a record is highlighted in a table, its corresponding feature is highlighted on the map. A highlighted record or feature is said to be selected.

10 Click the gray tab at the left edge of the first record in the table. The record is selected.

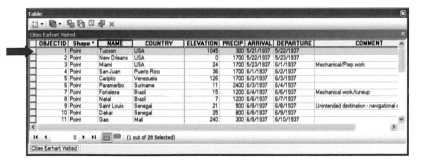

11 Move the attribute table away from the map display. Tucson is highlighted on the map. (The attribute table can be docked; however, for this exercise, do not dock it.)

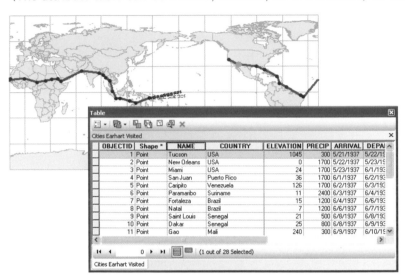

12 At the top of the attribute table, click the Table Options menu and click Clear Selection.

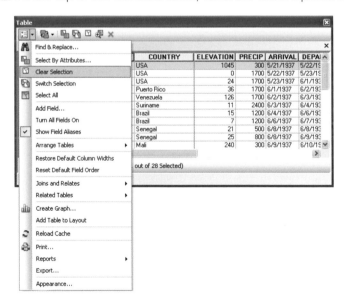

The record is unselected in the table, and the feature is unselected on the map.

13 In the table of contents, right-click the Flight Path layer and click Open Attribute Table.

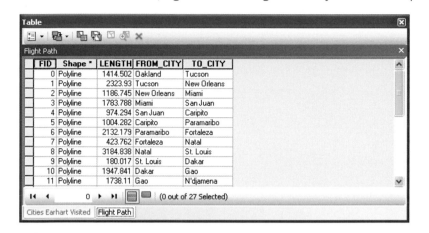

The attribute table opens. Notice the tabs at the bottom of the Table window. These tabs allow you to quickly switch between different attribute tables.

The Flight Path table contains a record for each stage of Earhart's flight. The attributes include the starting city (FROM_CITY), the destination city (TO_CITY), and the flight length (LENGTH) in kilometers.

The flight from Lae to Howland Island would have been 4,120 kilometers. The cumulative effect of small navigation errors over this distance might plausibly account for Earhart and Noonan's going well off course. You'll sort the LENGTH field to compare the distances of the flight segments they completed.

14 Right-click the LENGTH field name and click Sort Descending.

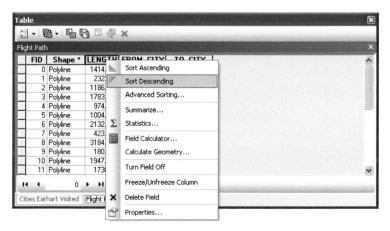

The records are ordered by length of flight segment from longest to shortest.

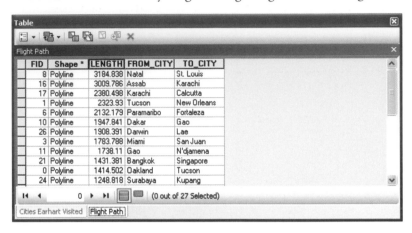

The longest completed leg of the flight, 3,184 kilometers, was over water from Natal, Brazil, to Saint Louis, Senegal. This was also a flight that had significant navigational error. (The aviators were about 175 kilometers off course from their intended destination, Dakar.)

Sorting a field is useful for seeing high and low values, but ArcMap can give you more detailed information.

15 Right-click the LENGTH field and click Statistics. The Statistics of flight_path window opens.

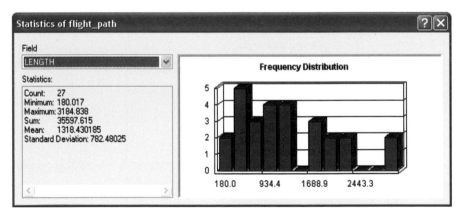

The Statistics box displays the number of records in the table (27) and the minimum, maximum, sum, mean, and standard deviation values. The average flight length, for example, was 1,318 kilometers.

The Frequency Distribution chart represents the distribution of values graphically. You can see that most of the flights were less than 1,500 kilometers and that only two were more than 2,500 kilometers. (The last bar in the chart reaches the 2 mark.)

16 Close the Statistics window. Close the Attributes of Flight Path table.

In the next chapter, you will see how the data for the Earhart project is managed in ArcCatalog.

17 If you want to save your work, click the File menu and click Save As. Navigate to **\GTKArcGIS\Chapter03\MyData**. Rename the file **my_ex03c.mxd** and click Save.

18 Click the File menu and click Exit. Click No if prompted to save your changes.

Chapter 4

Exploring ArcCatalog

Browsing map data
Searching for map data
Adding data to ArcMap

ArcCatalog is an application for managing geographic data. You can copy, move, and delete data; search for data; look at data before deciding whether to add it to a map; and create new data. The ArcCatalog application window includes the catalog display for looking at spatial data, the catalog tree for browsing data, and several toolbars.

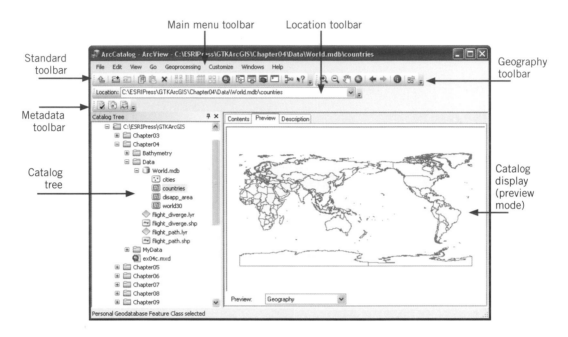

Spatial data comes in many different formats, including geodatabases, coverages, shapefiles, CAD (computer-aided design) files, rasters, and TINs (triangulated irregular networks). Each format is identified by its own icon in ArcCatalog. The shapefile icon, for example, is a green rectangle. Different patterns on the green rectangle distinguish point, line, and polygon shapefiles.

In this book, you'll use geodatabase, shapefile, and raster data. You'll also use layer files, which are not spatial datasets, but rather instructions for displaying spatial datasets with certain colors, symbol markers, line widths, and so on. Layer files, too, have their own ArcCatalog icon. You'll learn more about layer files and spatial data formats throughout this book.

Browsing map data

ArcCatalog gives you more information about spatial data than you can get from Windows Explorer or other file browsers. It can show you which folders contain spatial data and what kinds of spatial data they contain. It lets you preview features and attributes of data before you add the data to a map. It lets you examine and edit metadata, which is information about your data, such as when and how it was created.

Exercise 4a

As the GIS manager of the aviation history foundation's Earhart project, you need to be familiar with its spatial data. You'll use ArcCatalog to look at this data and get information about it.

The exercise instructions assume you have installed the data for *Getting to Know ArcGIS Desktop* to the default directory (*C:\ESRIPress\GTKArcGIS*). If you have installed the data elsewhere, you'll need to substitute the correct paths.

1 To open ArcCatalog, click the Start button on the Windows taskbar, point to All Programs, point to ArcGIS 10, and click ArcCatalog 10.

The ArcCatalog application window opens. The Catalog Tree lists the data and services that ArcCatalog is connected to. This is where connections to local drives on your computer are made. Your application may look different from the following graphic depending on the drives you have connected.

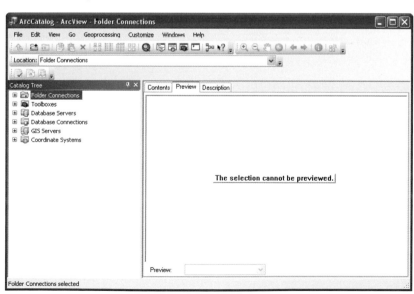

You can also connect to subdirectories, network drives, databases, Internet servers, and other services.

Once you connect to a folder, you can access the data it contains. In this exercise, it will be helpful to see the file extensions of different spatial datasets, so you will make sure that ArcCatalog is set to display these extensions.

2 Click the Customize menu and click ArcCatalog Options. In the ArcCatalog Options dialog box, click the General tab.

The ArcCatalog Options dialog box lets you specify the types of data ArcCatalog displays and the information it shows about them (file name, file size, date modified, and so on). You can distinguish folders containing spatial data from those that don't, and make many other customizations to the way data is displayed.

3 If necessary, uncheck Hide file extensions, then click OK.

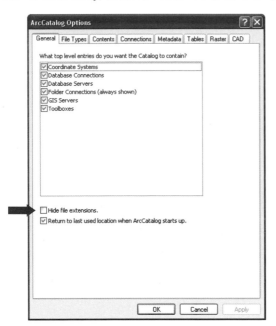

To access the spatial data in the GTKArcGIS folder more quickly, you'll create a connection to it.

4 On the Standard toolbar, click the Connect to Folder button.

The Connect to Folder dialog box opens. Your dialog box may look different depending on your local and network drives.

5 In the Connect to Folder dialog box, click the plus sign (+) next to the Local Disk (C:) drive to view its contents. Click the plus sign next to the **ESRIPress** folder to expand it. Click the **GTKArcGIS** folder as shown in the following graphic, then click OK.

A connection is made to C:\ESRIPress\GTKArcGIS.

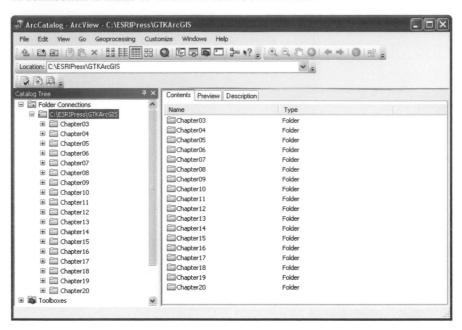

6 In the catalog tree, click the plus sign next to **C:\ESRIPress\GTKArcGIS**. Click the plus
sign next to the Chapter04 folder. It contains three folders, Bathymetry, Data, and
MyData, and one map document, **ex04c.mxd**.

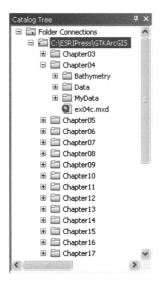

7 Click the plus sign next to the **Data** folder. It contains a geodatabase (World.mdb), two
shapefiles, and two layer files. Click the plus sign next to World.mdb.

Geodatabase
Feature class
Layer file
Shapefile

The geodatabase contains four feature classes. A feature class is a group of points, lines, or polygons representing geographic objects of the same kind. The cities feature class contains point features, the other three contain polygon features. This data was used to make the map of Earhart's flight in the previous chapter. For a more detailed explanation of feature classes, see the introduction to chapter 14.

8 On the Standard toolbar, make sure the Details button is selected.

9 In the catalog tree, click **World.mdb**. In the catalog display, make sure the Contents tab is active.

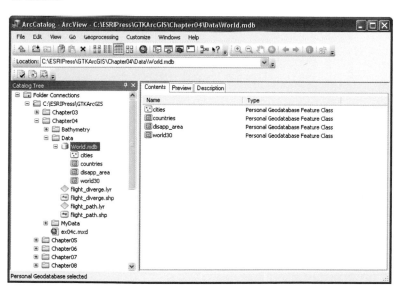

Four adjacent buttons on the Standard toolbar change how files look on the Contents tab. The Large Icons button displays large icons horizontally. The List button displays small icons vertically. The Details button is like the List button except that it also shows the file type—in this case, geodatabase feature classes. The Thumbnails button allows you to view small images of spatial datasets. Whenever a folder or a geodatabase is highlighted in the catalog tree, these four buttons are enabled.

10 On the Standard toolbar, click the Thumbnails button.

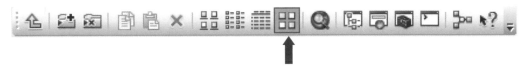

The display changes to show thumbnail images for each file. At the moment, the thumbnails are simply larger versions of the icons. A thumbnail that shows an image of a dataset can help you decide quickly if you want to use the data or not. You'll create a thumbnail for the countries feature class.

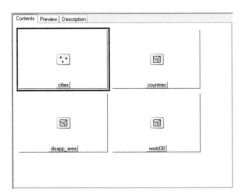

11 In the catalog tree, click the countries feature class in the World geodatabase. In the catalog display, click the Preview tab. The geographic data for **countries** displays in pale yellow.

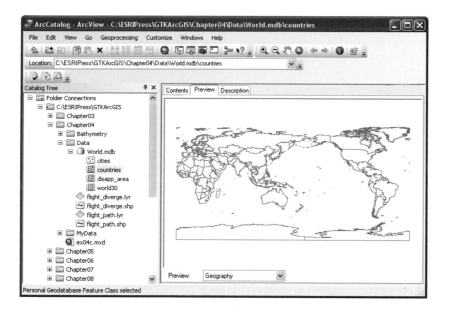

12 On the Geography toolbar, click the Create Thumbnail button.

An image of the data shown on the Preview tab is saved as a thumbnail graphic.

13 In the catalog display, click the Contents tab to see the thumbnail.

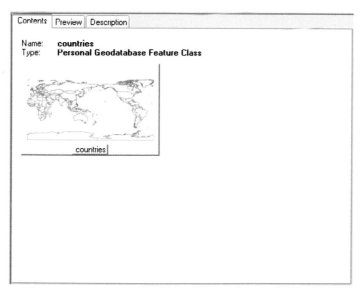

A thumbnail graphic may not be enough. The preview tab and the tools on the Geography toolbar let you further investigate a dataset before deciding if you want to use it in a map.

14 In the catalog display, click the Preview tab again. On the Geography toolbar, click the Zoom In tool.

15 Drag a box around the area north of Australia, as shown in the following graphic.

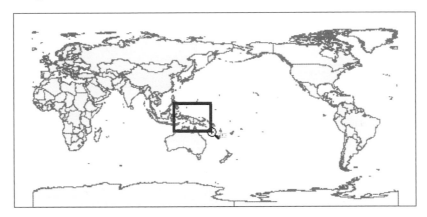

16 On the Geography toolbar, click the Identify tool.

17 Click a feature to identify it. If you don't see it flash green, move the Identify Results window away from the display and click the feature again.

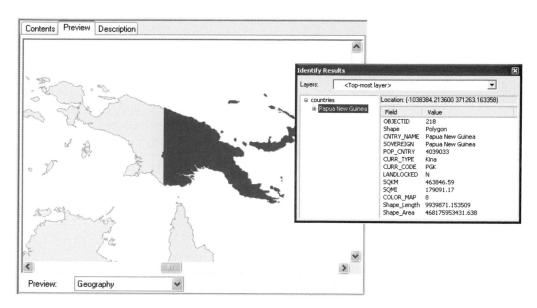

18 Close the Identify Results window. On the Geography toolbar, click the Full Extent button to zoom to the full extent of the data.

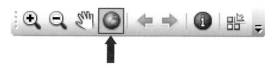

You can preview the attributes of a dataset as well as its geography.

19 At the bottom of the catalog display, click the Preview drop-down arrow and click Table. The display shows you the attribute table of the countries feature class.

OBJECTID *	Shape *	CNTRY_NAME	SOVEREIGN	POP_CNTRY
1	Polygon	Russia	Russia	151827600
2	Polygon	Svalbard	Norway	3148
3	Polygon	Greenland	Denmark	55413
4	Polygon	Jan Mayen	Norway	-99999
5	Polygon	Iceland	Iceland	267240
6	Polygon	Faroe Is.	Denmark	47067
7	Polygon	Finland	Finland	5031379
8	Polygon	Estonia	Estonia	1590808
9	Polygon	Latvia	Latvia	2690291
10	Polygon	United Kingdom	United Kingdom	56420180
11	Polygon	Lithuania	Lithuania	3786560
12	Polygon	Denmark	Denmark	4667750
13	Polygon	Belarus	Belarus	10521400
14	Polygon	Isle of Man	United Kingdom	71296
15	Polygon	Ireland	Ireland	5015975
16	Polygon	Poland	Poland	37911870

1 ▸ ▸| (of 252)

Preview: Table

20 Click the Preview drop-down arrow again and click Geography.

The third tab in the catalog display is the Description tab. Metadata provides information about a dataset, such as its coordinate system, its spatial extent, and descriptions of its attributes. It may also explain how and when the data was created, what standards of accuracy it meets, and what its appropriate uses are.

A great deal of metadata is maintained automatically by ArcCatalog; some, however, must be maintained by the people who use and manage the data. ArcCatalog doesn't require you to maintain metadata, but you should.

21 In the catalog display, click the Description tab. If needed, scroll down to see the thumbnail you created, tag words, summary, description, and other information.

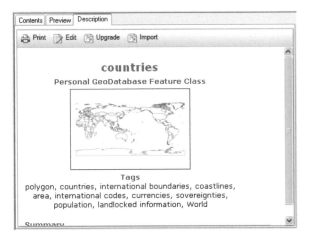

22 Click the Edit button inside the description to access metadata about the countries feature class.

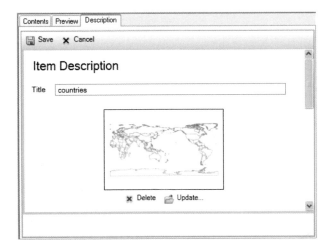

Metadata can be generated inside the Description tab. The format of metadata has changed with ArcGIS 10. All metadata created in prior releases must be upgraded to the current ArcGIS metadata format. This upgrade will not remove any existing metadata—existing metadata can be accessed in the same manner as before with previous releases of ArcGIS. You can include useful details about the file including spatial reference information and attribute information. Any of the information here can be edited. However, it will not be done for this exercise.

23 Browse the different categories to see what type of metadata could be entered.

For more information about metadata, click the For more information about metadata, click the Contents tab in ArcGIS Desktop Help (not the Contents tab in ArcCatalog) and navigate to *Professional Library > Data management > Geographic data types > Metadata.*

24 In the catalog display, click the Contents tab. In the catalog tree, click **World.mdb**. On the Standard toolbar, click the Details button.

The display is restored to its initial state.

25 If you are continuing with the next exercise, leave ArcCatalog open. Otherwise, exit the application.

Searching for map data

With ArcCatalog, you can search for geographic data on disk or across a network. You can search by name, file type, geographic location, creation date, or by other properties stored in the metadata.

Exercise 4b

In this exercise, you will search ArcCatalog for more data that may be useful to the Earhart project.

1 If necessary, start ArcCatalog.

If ArcCatalog is open from the previous exercise, you are connected to the **C:\ESRIPress\GTKArcGIS** folder, as shown in the following graphic.

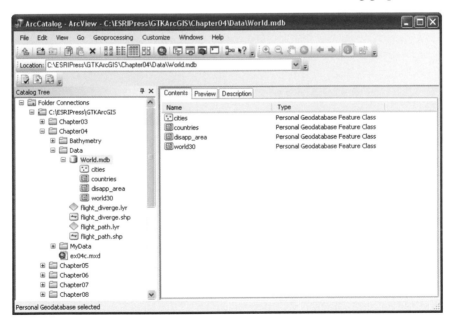

If you started a new ArcCatalog session, you may be connected to another folder and your catalog tree may look slightly different. If needed, browse to the Data folder listed above.

2 On the Standard toolbar, click the Search button.

The Search window opens. Depending on how your interface is set up, it may be docked to the right side of the ArcCatalog interface or it may be floating. You can work with it in either docked or floating orientation.

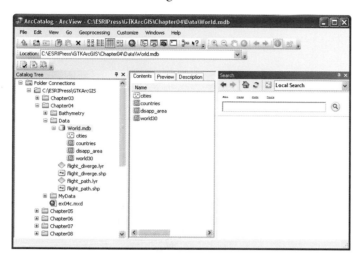

The Search window gives you more flexibility to search for maps, data and tools. Data and tools from a search result can be used with your active map or model. It indexes data so that it can be searched more efficiently and quickly.

Before ArcCatalog can search for maps or data, you must specify a search location—such as a folder or geodatabase. An index of the contents of the specified search location will be created by ArcCatalog. As you will see in a moment, the search results are listed in the Search window as shortcuts, or pointers, to the maps, data and tools it finds.

You will define the search to the contents of the Chapter04 folder. This spares ArcCatalog the trouble of looking for data in unlikely places.

3 In the Search window, click the Index/Search Options button.

The Index/Search Options dialog box opens.

4   In the Register Folders and Server Connections frame, click the Add button. In the Browse Folders to be Indexed dialog box, navigate to **C:\ESRIPress\GTKArcGIS**. Click the **Chapter04** folder, as shown in the following graphic, then click Select.

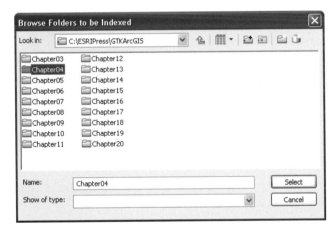

The search is now limited to the contents of the Chapter04 folder.

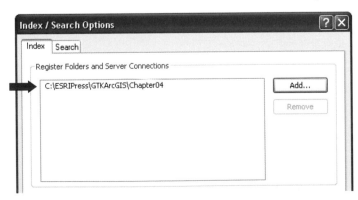

However, before the Chapter04 folder can be searched properly, it must be indexed.

5 Click the Index New Items button.

In a few seconds, the content of Chapter04 is indexed.

Indexing Status	
Items Indexed:	12
Indexing Status:	Active
Last Indexing Start Time:	3/24/2010 4:41 PM
Last Indexing Duration:	2 minutes 20 seconds
Next Indexing Start Time:	

Twelve items are indexed and the time and duration of the index are recorded. Indexing duration depends on the performance of your system and the number of folder or server connections to be indexed. The Index/Search Options dialog box may be closed at any time and will not interrupt indexing.

6 Click OK to close the Index/Search Options dialog box.

You will concentrate your search for data about Earhart using keywords stored in a dataset's spatial metadata. If many matches are found, the dataset and its description are returned in the search results.

7 In the Search window, type **Earhart** into the search box and click the search button adjacent to it.

Search results return with all data that have the keyword "Earhart" in its metadata. Hover over a search result to view a brief description of it. However, we are interested in the ocean area where Earhart disappeared.

8 In the search box, delete "Earhart," type **ocean**, and click the search button.

The Search window lists three datasets that were found. There is a raster dataset called seafloor that you may have not seen before (at the top of the list.)

9 In the top search result listing, click the text **seafloor** (or the path link below it.) Make sure you click the text link and not the description. Clicking the description opens the Item Description window.

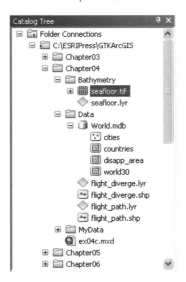

The catalog tree opens to the location of the data. If necessary, scroll to the top of the table of contents to see it.

10 Close the Search window.

The seafloor raster dataset is in the Bathymetry folder under Chapter04, along with a layer file called **seafloor.lyr**.

11 In the catalog tree, make sure that **seafloor.tif** is highlighted. In the catalog display, click the Preview tab.

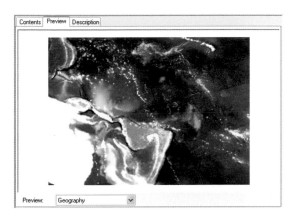

You see a raster dataset of seafloor elevation in East Oceania. The data may be useful in a map of the area where Earhart disappeared. For example, you could zoom in on Nikumaroro Island and find out how deep the water around it is. This information might affect operations to find and recover plane wreckage. The black-and-white image, however, is not easy to interpret as seafloor elevation. The seafloor layer file displays the elevation data in shades of blue.

12 In the catalog tree, click **seafloor.lyr**.

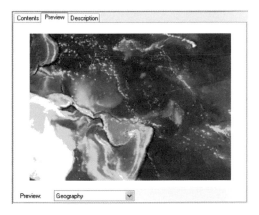

In the layer file, the depth of the water is indicated by the shade of blue; darker is deeper. White areas are land.

13 On the Geography toolbar, click the Identify tool.

14 Click a few locations to identify elevation values (described as pixel values in the Identify Results window). The values are in meters. If you click a spot of land, you should get the value No Data.

15 Close the Identify Results window. In the catalog tree, click the minus sign next to the Bathymetry folder to collapse it. If necessary, click the plus sign next to the Data folder to expand it.

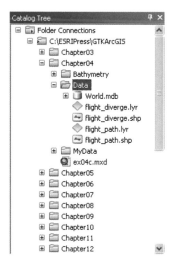

Besides the World geodatabase, the Data folder contains shapefiles and layer files for the Earhart flight paths.

16 Make sure the Preview tab is active in the display. Click the flight_diverge shapefile (green square icon) to preview it. Next, click the flight_diverge layer file (yellow diamond) and preview it.

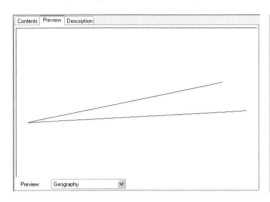

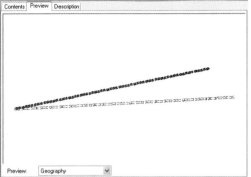

flight_diverge shapefile   flight_diverge layer file

The shapefile displays as two blue lines. The layer file shows the thick dotted lines that you saw in the previous chapter.

Layer files allow you to store symbology information—the colors, shapes, and sizes that you choose for features—so you never have to re-create it. Every layer file is associated with and depends on a spatial dataset. The flight_diverge layer file won't display in ArcMap unless the flight_diverge shapefile is accessible on disk.

In the next exercise, you will add data and layer files from ArcCatalog to ArcMap.

17 If you are continuing with the next exercise, leave ArcCatalog open. Otherwise, exit the application.

Adding data to ArcMap

One way to add data to ArcMap is to drag it from ArcCatalog. Once it's there, you can look at it, as you did in the previous chapter, as a map display and table of contents. This is called data view. You can also look at it as if it were on a page that you send to a printer. This is called layout view.

In layout view, you see map layers organized in one or more rectangles on a larger background rectangle. The smaller rectangles are called data frames and the background is called the virtual page.

A map document can have one data frame or many. Data frames may contain different sets of data (usually related to a common subject) or they may show different views of the same data.

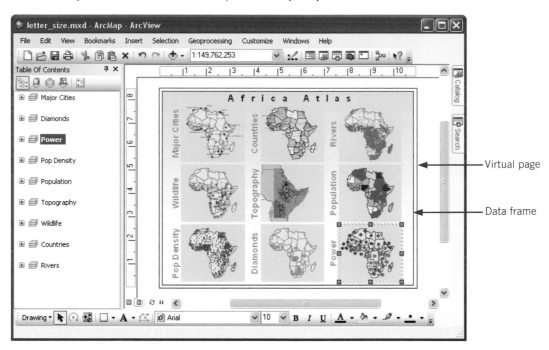

In data view, you see just one data frame at a time.

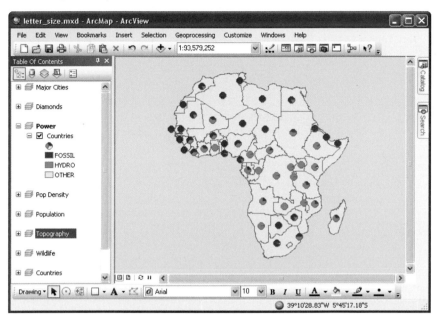

The display window shows the contents of the active data frame (in this case, power and energy data). The name of the active data frame is boldfaced in the table of contents.

Exercise 4c

The aviation history foundation hopes to raise money for an expedition to look for the wreckage of Earhart's plane. Part of your job is to create maps to interest potential investors. In chapter 19, you'll learn in detail about creating maps for presentation. For now, you'll learn how to add data to ArcMap from ArcCatalog and how to work with data frames in ArcMap.

1   If necessary, start ArcCatalog.

If ArcCatalog is open from the previous exercise, you are connected to **C:\ESRIPress \GTKArcGIS\Chapter04\Data\flight_diverge.lyr**, as shown in the graphic on the next page. If you started a new ArcCatalog session, navigate to this location now and click the Preview tab if necessary.

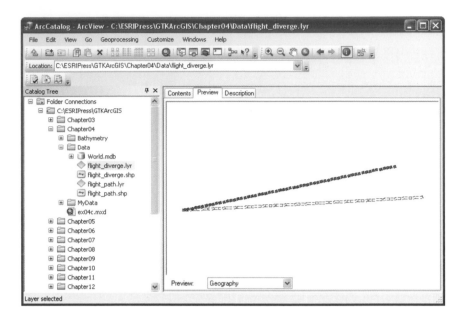

2   On the Standard toolbar, click the Launch ArcMap button.

3   In the ArcMap—Getting Started dialog box, click "Browse for more..." to start using ArcMap with an existing map. In the Open file browser, navigate to **C:\GTKArcGIS \Chapter04**. Click the file **ex04c.mxd** to highlight it and click Open.

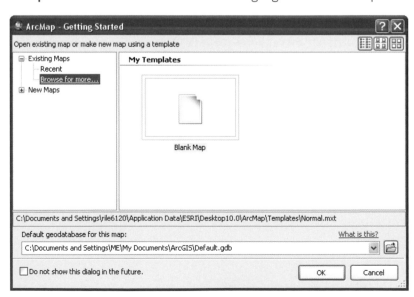

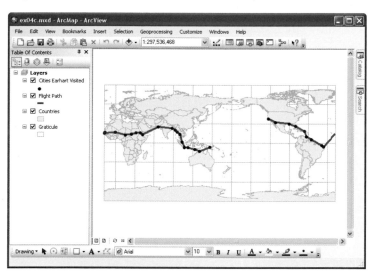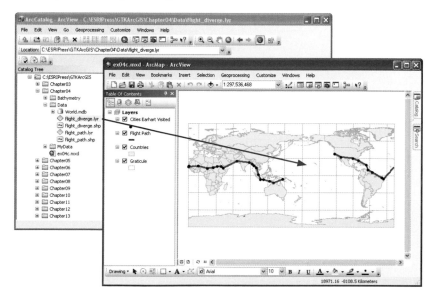

4   A preview of ex04c.mxd is shown. Click Open.

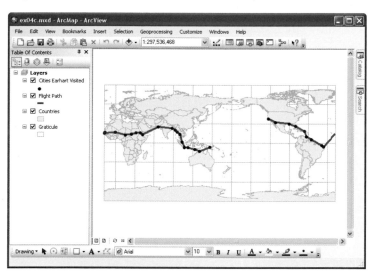

You see the familiar map of Amelia Earhart's flight. The map has one data frame that contains four layers. You will add a layer for the diverging flight paths.

5   Position the ArcCatalog and ArcMap application windows so that you can see both the catalog tree and the ArcMap display window.

6   In the catalog tree, click the **flight_diverge** layer file (yellow icon). Drag the file anywhere over the ArcMap display window and release the mouse button. Bring ArcMap forward by clicking its title bar.

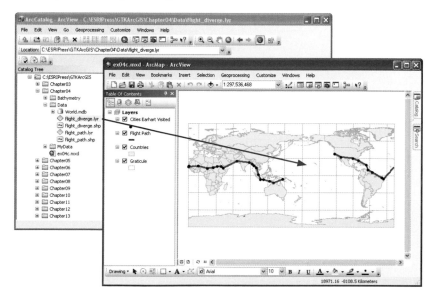

The layer file is added to ArcMap.

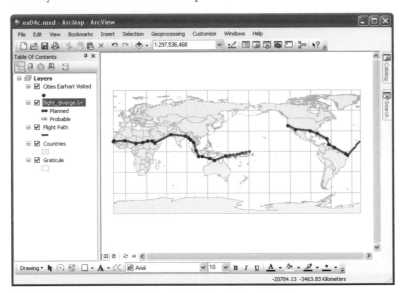

You'll rename the layer file.

7   In the table of contents, right-click flight_diverge.lyr and click Properties.

8   In the Layer Properties dialog box, click the General tab. Replace the layer name with **Diverging Flight Paths**, as shown in the following graphic, then click OK.

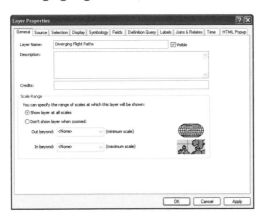

The layer name is changed in the table of contents.

9   In ArcMap, click the View menu and click Layout View.

You see the map as if it is on a piece of paper. A new toolbar, the Layout toolbar, opens. It can be docked or left floating.

The map document has a single data frame, occupying the top half of the virtual page. Some other elements have been added to the layout for you. There is a title above the data frame, a green rectangle graphic that marks the area where Earhart disappeared, and another title that will accompany a data frame you are about to add. This new data frame will show a zoomed-in view of the area of disappearance.

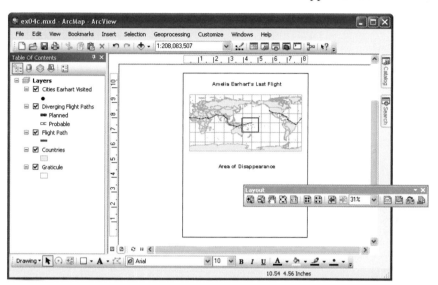

10 Click the Insert menu and click Data Frame.

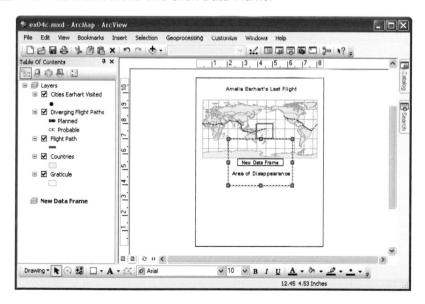

A data frame is added to the layout and its name, New Data Frame, appears in the table of contents. The name is boldfaced, indicating that it is the active data frame—the one that layers will be added to.

11 In the layout window, move the mouse pointer over the new data frame. The cursor changes to a four-headed arrow. Drag the frame beneath the "Area of Disappearance" title.

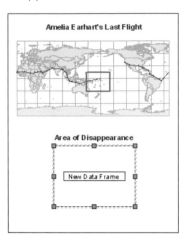

The name of a data frame is independent of any title you may choose to add to the layout. By default, the first data frame in a map document is called "Layers" and the second is called "New Data Frame." It is often helpful to rename a data frame to something more descriptive (and you will do that later in this exercise), but it is not necessary.

You'll switch back to data view to add layers to the new data frame. (You could do this in layout view as well.)

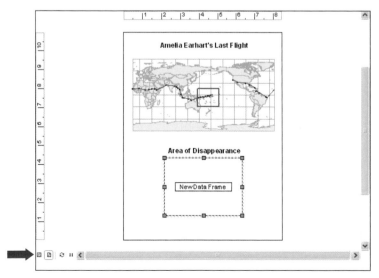

12 Click the Data View button in the lower left corner of the layout window. This button and the Layout View button next to it can be used instead of the View menu to switch between views.

The display window is empty because the active data frame (New Data Frame) contains no layers.

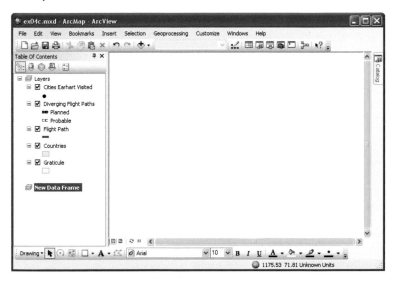

13 Make ArcCatalog the active application by clicking its title bar. In the catalog tree, click the plus sign next to the World geodatabase and click **disapp_area**. Drag the feature class to the ArcMap display window and release the mouse button.

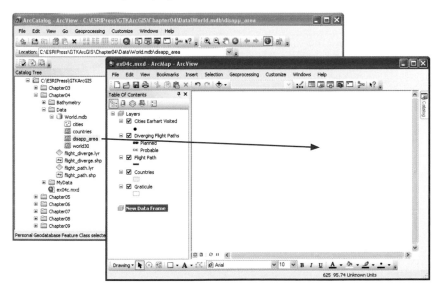

14 Bring ArcMap forward by clicking its title bar. The feature class is added as a layer to the new data frame.

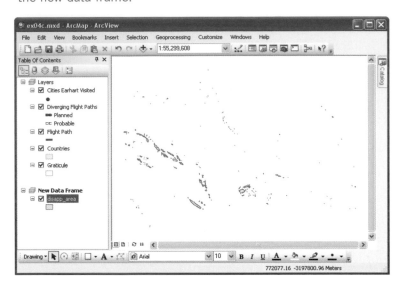

ArcMap assigns a random color to the disapp_area layer, so your color may be different.

15 Make ArcCatalog the active application. In the catalog tree, click the plus sign by the Bathymetry folder and click **seafloor.lyr** (yellow icon). Drag the layer file to ArcMap and bring ArcMap forward.

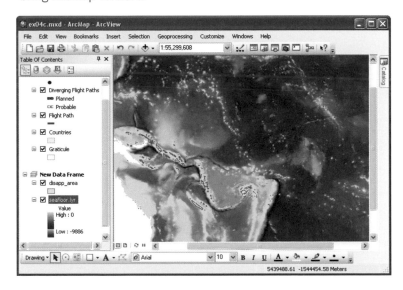

The seafloor.lyr layer displays underneath the disapp_area layer.

Now you will copy the Diverging Flight Paths layer from the Layers data frame to the new data frame.

16 In the ArcMap table of contents, right-click the Diverging Flight Paths layer in the Layers data frame. On the context menu, click Copy.

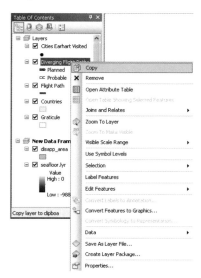

17 In the table of contents, right-click New Data Frame. On the context menu, click Paste Layer(s).

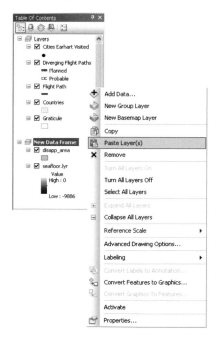

The Diverging Flight Paths layer is pasted into the new data frame and displays on the map.

18 Click the Layout View button in the lower left corner of the layout window.

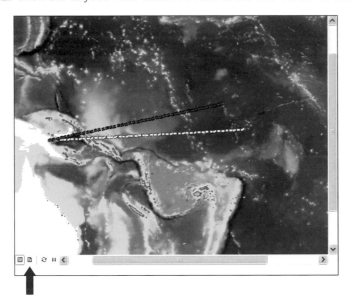

The data frame displays the data you've added to it.

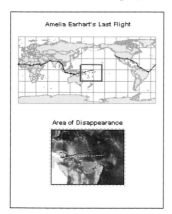

Now you'll change the color of the disapp_area layer to make it contrast with the blue of the seafloor layer.

19 Click the Data View button to switch back to data view. In the table of contents, right-click the polygon symbol for the disapp_area layer. A color palette opens.

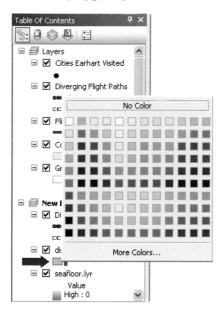

20 Move the mouse pointer over the color palette. Each color square is identified by name. Click Sahara Sand.

The color of the layer is updated in the table of contents and in the map features.

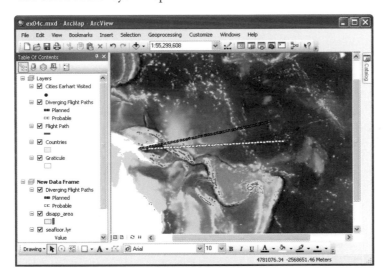

Finally, you will rename the data frame.

21 In the table of contents, right-click New Data Frame. On the context menu, click Properties. The Data Frame Properties dialog box opens. Click the General tab.

22 In the Name box, New Data Frame is highlighted. Type **Area of Disappearance** in its place, as shown in the following graphic, then click OK.

The new name is displayed in the table of contents.

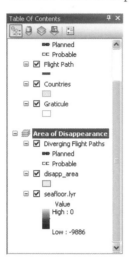

You know how to preview data in ArcCatalog and look at its metadata. You can use ArcCatalog to search for spatial data on disk. And you know how to add layers to ArcMap from ArcCatalog. You'll have no trouble managing the data for the Earhart project.

23 If you want to explore more in ArcCatalog, leave it open. Otherwise, exit the application.

24 If you want to save your work, click the File menu in ArcMap and click Save As. Navigate to **\GTKArcGIS\Chapter04\MyData**. Rename the file **my_ex04c.mxd** and click Save.

25 Close the Layout toolbar. If you are continuing to the next chapter, leave ArcMap open. Otherwise, exit the application. Click No when prompted to save changes.

Chapter 5

Symbolizing features and rasters

Changing symbology
Symbolizing features by categorical attributes
Using styles and creating layer files
Symbolizing rasters

Symbolizing features means assigning them colors, markers, sizes, widths, angles, patterns, transparency, and other properties by which they can be recognized on a map. Symbols often look like the objects they represent, as when a lake polygon is blue and fire hydrants are marked by icons that look like fire hydrants. Sometimes the relationship is less straightforward. On a street map, varying line thicknesses may show whether a street is a local road, an arterial, or a highway without implying that the widths on the map are proportional to the widths of the actual streets. Symbology may also be purely conventional. Cities, for example, are commonly symbolized as circles, though cities are seldom round.

By varying symbol properties, you convey information about features.

Left: Fill patterns and colors differentiate intermittent lakes, perennial lakes, and salt pans. Center: Line thickness, color, and solidity show railways and main, secondary, and local roads. Right: Unique icons mark major and minor cities.

Symbology is also influenced by scale. A city may be a circle on one map and a polygon on another.

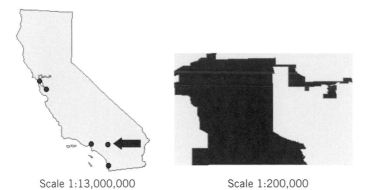

Scale 1:13,000,000   Scale 1:200,000

Left: Redlands, California, represented as a circle. Right: Redlands, California, represented as a polygon.

ArcGIS has thousands of symbols for common map features. Symbols are organized by style (Environmental, Transportation, Weather, and so on) to make it easy to locate the ones you need. You can also create your own symbols.

When a dataset is added to ArcMap, all its features have the same symbology. To assign new symbology, you use information from a field in the layer attribute table.

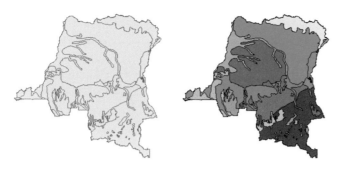

ID	Vegetation Type
0	Albertine Rift Forest
1	Angolan Miombo Woodland
2	C Congolian Low. Forests
3	Central Africa Mangroves
4	Congolian Coastal Forests
5	Congolian Riverine Forest
6	Congolian Swamp Forests
7	East Sudanian Savanna
8	Itombwe Montane Forests
9	Marungu Montane Forest
10	N. Kal. Sand Grasslands
11	NE Congolian FSM
12	NE Congolian Low. Forests
13	NW Congolian Low. Forests
14	Rwenzori Virunga Moorland
15	S Kal. Sands Grasslands
16	Sth Congolian FSM

Ecoregions in the Democratic Republic of the Congo. Left: All features have the same symbology. Center: Each feature has a different fill color. Right: The Vegetation Type values symbolized by the fill colors.

You can save a layer's symbology by making a layer (.lyr) file. When you add a layer file to a map, the features are already symbolized the way you want.

Symbolizing raster layers is similar to symbolizing feature layers, but with fewer options. The only property that can be controlled is the color of the cells that comprise the raster.

Changing symbology

Datasets added to ArcMap have default symbology. Points are displayed with small circles, for instance, and polygons have outlines. The colors for points, lines, and polygons are randomly chosen.

In this exercise, you'll change the default symbology for a polygon layer of countries and a point layer of cities. You'll also change the background color of the data frame.

Exercise 5a

You are a graphic designer donating your time to create a poster about Africa for Geography Awareness Week. Sponsored by the National Geographic Society, Geography Awareness Week is held every November and includes GIS Day among its many events and programs.

The Africa Atlas poster will consist of nine maps displaying cities, countries, rivers, wildlife, topography (surface relief), population, and natural resources. It will be distributed to elementary schools around the country.

The completed poster, measuring 44 by 34 inches, will look like the one in the following graphic. In this exercise, you will create the first of the nine maps, depicting major cities.

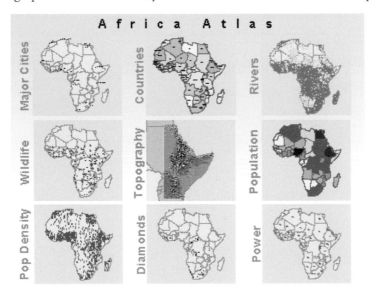

When you finish chapter 6, you can send your poster to a plotter, if you have access to one. Alternatively, you can send a letter-size version of the poster to a desktop printer.

1   Start ArcMap. In the ArcMap—Getting Started dialog box, under the Existing Maps section, click Browse for more. (If ArcMap is already running, click the File menu and click Open.) Navigate to **C:\ESRIPress\GTKArcGIS\Chapter05**. Click **ex05a.mxd** and click Open.

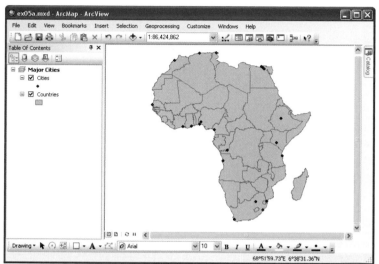

The table of contents has a data frame called Major Cities that includes a Cities layer and a Countries layer. Countries are symbolized in pale purple, cities as dark gray points. Each city has a population of one million or more.

You'll change the symbology for both the countries and the cities.

2   In the table of contents, right-click the symbol for the Countries layer to open the color palette. In the palette, click Sahara Sand.

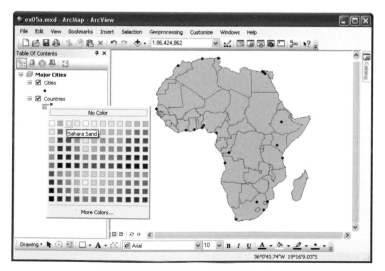

The countries are redrawn in the new color. Polygon features are composed of two symbols, a fill and an outline. To change the color or width of the outline, you open the Symbol Selector.

3   In the table of contents, click the symbol for the Countries layer. The Symbol Selector dialog box opens.

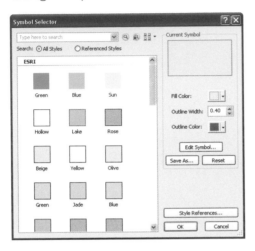

The scrolling box on the left contains predefined symbols. The Options frame on the right allows you to pick colors and set outline widths.

4   In the Options frame, click the Outline Color square to open the color palette. In the color palette, click Gray 40%. Click OK at the bottom of the Symbol Selector dialog box.

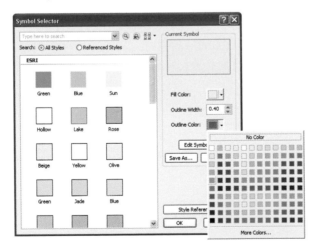

The cities stand out more distinctly against the lighter gray outlines of the countries.

Now, you'll change the symbology for the cities. You'll make them larger, change their color, and label them.

5 In the table of contents, click the point symbol for the Cities layer. The Symbol Selector dialog box opens. In the scrolling box of predefined point symbols, click Circle 2. In the Options frame, click the Size down arrow to change the symbol size to 10 points.

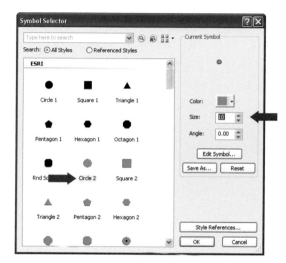

6 Click the Color square to open the color palette. In the color palette, click Ginger Pink. Click OK to close the Symbol Selector dialog box.

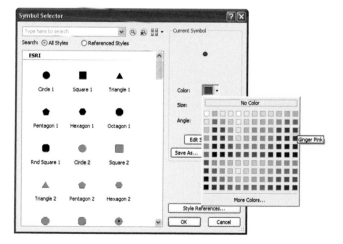

On the map, the cities display with the new symbol.

7   In the table of contents, right-click the Cities layer name (not the symbol) and click Label Features.

The cities are labeled with their names. Depending on the size of your ArcMap window, and the scale, your labels may be positioned differently from those in the graphic. In chapter 7, you'll learn how to change the size, color, and font of labels.

Next, you'll change the background color of the data frame.

8 In the table of contents, right-click the Major Cities data frame, and click Properties. The Data Frame Properties dialog box opens. Click the Frame tab (not the Data Frame tab).

9 On the Frame tab, click the Background drop-down arrow and click Lt Blue. When the background color is applied, as shown in the following graphic, click OK to close the Data Frame Properties dialog box.

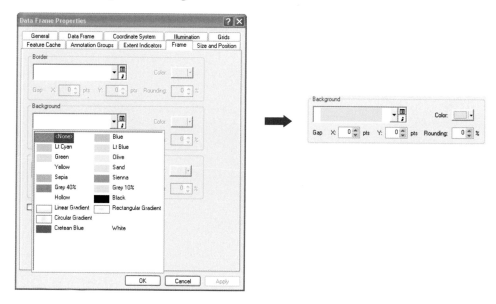

The color is applied to the data frame.

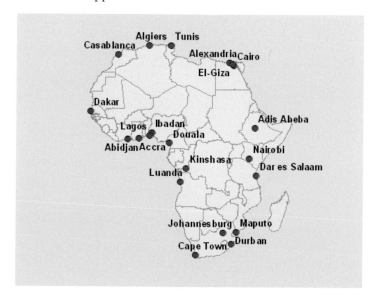

You have finished the major cities map. In the next exercise, you'll make a map in which each country has unique symbology. You'll also symbolize the rivers of Africa.

10 If you want to save your work, click the File menu and click Save As. Navigate to **\GTKArcGIS\Chapter05\MyData**. Rename the file **my_ex05a.mxd** and click Save.

11 If you are continuing with the next exercise, leave ArcMap open. Otherwise, exit the application. Click No if prompted to save your changes.

Symbolizing features by categorical attributes

In the previous exercise, each city had the same marker symbol and all countries were drawn in the same color. You can make symbology more informative by assigning a different symbol to each unique value, or to ranges of values, in the layer attribute table. For example, you might symbolize a polygon layer of countries according to their names. Every feature would have a unique color because every country has a different name. You might also symbolize countries by their political systems. Features with the same system would have the same color: republics might be blue, constitutional monarchies purple, and communist states red.

Attributes that are names or descriptions are called categorical attributes (or qualitative or descriptive attributes). They are usually text, but they may be numbers if the numbers are codes standing for descriptions. Attributes that are measurements or counts of features are called quantitative attributes. A country's area in square kilometers is a measurement. Its population is a count.

In this exercise, you'll symbolize features by categorical attributes. You'll work with quantitative attributes in chapter 6.

Exercise 5b

You are continuing your work on the Africa Atlas poster. You'll symbolize countries by name, so that each has a unique color. Then you'll symbolize African rivers. To give each river a unique color would be confusing—there are too many of them. Instead, you'll symbolize them by an attribute that identifies them as perennial or intermittent.

1   Start ArcMap. In the ArcMap—Getting Started dialog box, under the Existing Maps section, click Browse for more. (If ArcMap is already running, click the File menu and click Open.) Navigate to **C:\ESRIPress\GTKArcGIS\Chapter05**. Click **ex05b.mxd** and click Open.

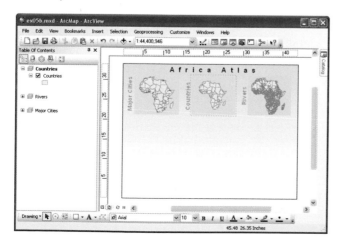

The map opens in layout view. You see the Major Cities data frame from the previous exercise and two new ones: Countries and Rivers. In the map display, the Countries data frame has a dashed line around it because it is active.

2   Click the View menu and click Data View. On the Tools toolbar, click the Full Extent button.

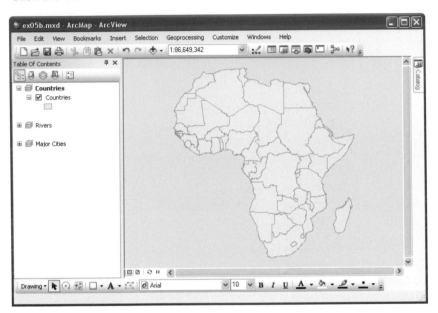

The Countries data frame is active and contains one layer, also called Countries. All features in the layer are symbolized with the same color. To give each feature its own color, you'll open the Layer Properties dialog box.

3   In the table of contents, right-click the Countries layer (not the Countries data frame) and click Properties. The Layer Properties dialog box opens. Click the Symbology tab.

The Show box on the left side of the dialog box lists different methods for symbolizing features. Some methods have more than one option. By default, Features is selected. It has only one option—Single symbol.

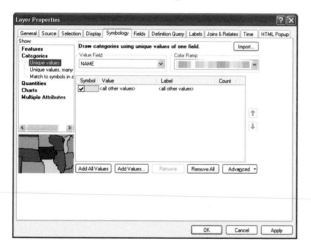

Since you want to symbolize each country with its own color, you need an attribute with a unique value for each feature—the country name would be a natural choice.

4 In the Show box, click Categories.

The Unique values option of the Categories method is highlighted and the dialog box changes. The Value Field drop-down list contains the fields in the attribute table that can be used to symbolize the countries.

The Color Scheme drop-down list contains different color patterns. If this is the first time you have set a color scheme in ArcMap, you should see a bar of pastel colors; otherwise, you'll see whichever scheme you applied last.

The large window in the middle of the dialog box shows which features in the layer are being symbolized. You can symbolize all features or just some of them.

5   Make sure the Value Field drop-down list is set to NAME. At the bottom of the dialog box, click Add All Values.

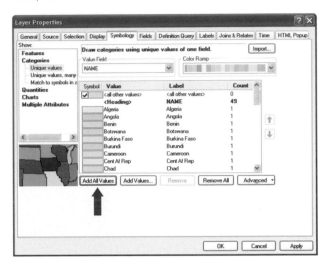

Each feature in the Countries layer is listed in the Value column and has a symbol from the color scheme assigned to it. The Label column displays the feature name as it will appear in the table of contents. (It will be the same as the value unless you change it.)

The Count column tells you the number of features being symbolized and the number of features that have each value. In this case, the count is one for each value because each country's name is unique.

6   Click the Color Ramp drop-down arrow to see the list of color schemes. Scroll up and click the scheme at the top of the list. The features are assigned new symbols from the Basic Random Scheme. (If this happens to be the scheme that is already set, choose a different one.)

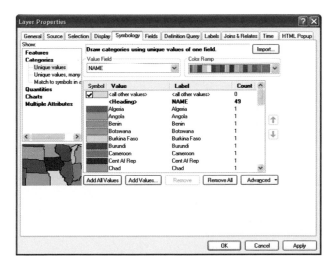

For this map, you'll go back to pastels.

7 In the Color Ramp drop-down list, right-click the colors (not the drop-down arrow). On the context menu, click Graphic View to uncheck it. The image of the color scheme is replaced by its name. Click the Color Ramp drop-down arrow and scroll back down to Pastels. Click to select it.

8 In the Symbol column, click the check box next to <all other values> to uncheck it. The <all other values> symbol is used when you want to assign unique symbology to some features in a layer and want the rest to be identical. That isn't the case here. Make sure your dialog box matches the following graphic, then click OK.

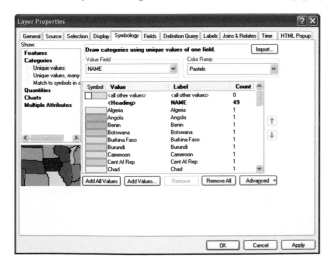

In the map, the countries are symbolized in pastels. (The colors of the countries may be different on your map.) The table of contents shows the name of each country and its symbol.

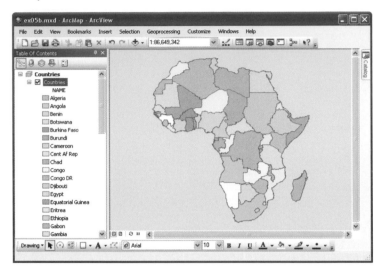

After a color scheme has been applied, you can change individual colors. For example, you could right-click the symbol for Algeria in the table of contents and change it to Mars Red or Steel Blue.

The countries are distinguished by unique symbols, but the symbols don't make it easy to identify countries. (How quickly can you find Malawi?) You will label the countries with their names.

9   In the table of contents, right-click the Countries layer and click Label Features.

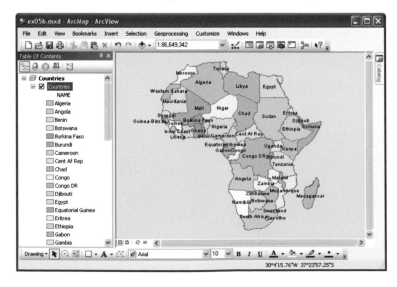

As with the cities, your country labels may look different depending on the size of your ArcMap window. Your country map is finished. Your last task in this exercise is to create a map of African rivers.

10 In the table of contents, click the minus signs next to the Countries layer and the Countries data frame.

11 Right-click the Rivers data frame and click Activate. Click the plus sign next to the data frame. If necessary, click the Full Extents button. The data frame contains a Rivers layer and a Countries layer.

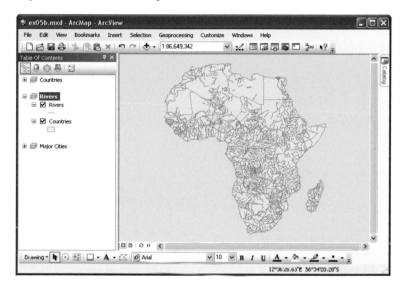

Before you symbolize the Rivers layer, you'll look at its attribute table to see what information is available.

12 Right-click the Rivers layer and click Open Attribute Table.

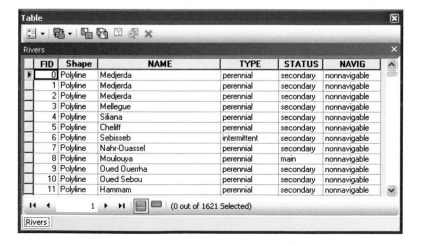

The attributes include the river name, its type (perennial or intermittent), its status (main or secondary), and its navigability. You'll symbolize rivers by type.

13 Close the table. In the table of contents, double-click the Rivers layer to open its Layer Properties. (Double-clicking is a shortcut for right-clicking and choosing Properties.) On the Layer Properties dialog box, click the Symbology tab.

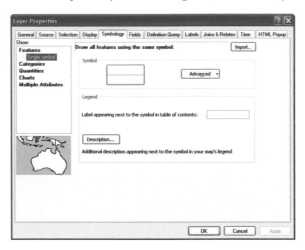

14 In the Show box, click Categories and make sure the Unique values option is highlighted. Click the Value Field drop-down arrow and click TYPE. Click Add All Values.

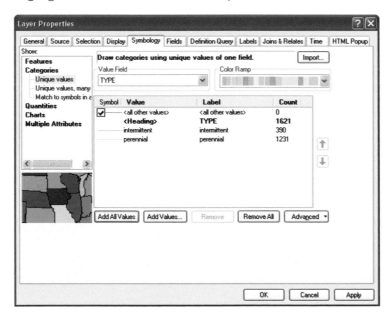

There are two values in the TYPE field: intermittent and perennial. At the moment, they are symbolized with pastels. You'll change the symbology.

Adding values

You may want to focus on some features in a layer and deemphasize or ignore others. The Add Values button (next to Add All Values) lets you select particular attribute values and assign symbology to features that have them. Other features are assigned a single symbol—the one designated for <for other values>. If the <all other values> check box is unchecked, these features are not displayed at all.

15 In the Symbol column, double-click the line symbol next to the intermittent value to open the Symbol Selector.

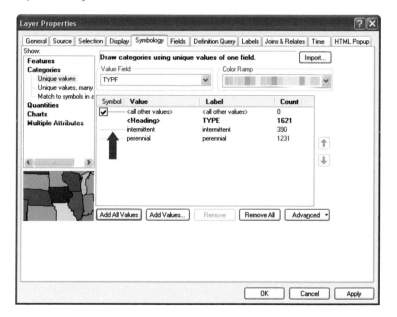

16 In the Symbol Selector, scroll down two thirds of the way until you see the Stream, Intermittent symbol. Click the symbol as shown in the following graphic, then click OK in the Symbol Selector.

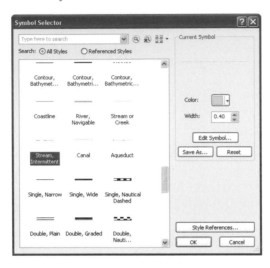

17 In the Symbol column of the Layer Properties dialog box, double-click the symbol next to the perennial value. The Symbol Selector opens again. This time, type **river** into the search box and click the search button adjacent to it. (It may take a minute for the symbols to index.)

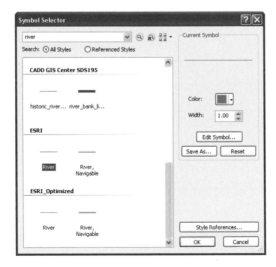

The symbol selector search utility allows you to quickly find a symbol.

18 Scroll down the list. Under the ESRI styles, click the River symbol, then click OK.

19 In the Symbol column of the Layer Properties dialog box, click the check box next to <all other values> to uncheck it. Make sure that your dialog box matches the following graphic, then click OK.

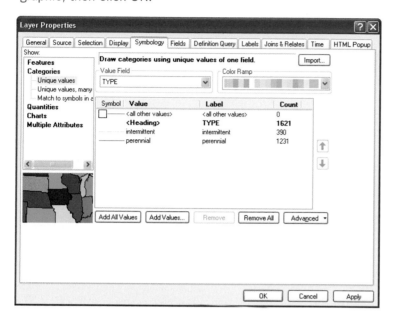

In the map, the rivers are displayed with the new symbology.

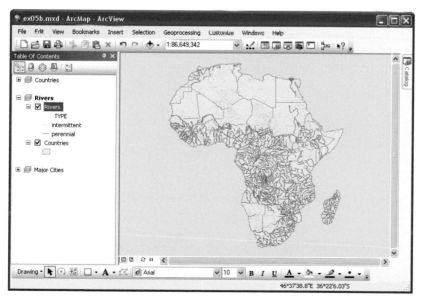

The map shows that central Africa is a dense network of perennial rivers while the northern and southwestern parts of the continent have few rivers that run year-round.

You'll check your progress on the Africa Atlas poster.

20 Click the View menu and click Layout View.

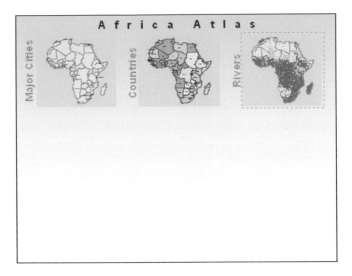

Your layout has three data frames, showing major cities, countries, and rivers. In the next exercise, you'll symbolize wildlife habitat.

21 If you want to save your work, click the File menu and click Save As. Navigate to **\GTKArcGIS\Chapter05\MyData**. Rename the file **my_ex05b.mxd** and click Save.

22 If you are continuing with the next exercise, leave ArcMap open. Otherwise, exit the application. Click No if prompted to save your changes.

Using styles and creating layer files

In the previous exercises, you worked with point, line, and polygon symbols in the Symbol Selector dialog box, colors on the color palette, color schemes in the Layer Properties dialog box, and background colors in the Data Frame Properties dialog box.

All these symbols and colors belong to the ESRI style. A style is a collection of predefined symbols, colors, and other map elements such as labels, north arrows, scale bars, and borders. ArcGIS has over twenty styles, and you can make new ones by combining elements from existing styles and creating your own symbols.

Conservation style

Public Signs style

Hazmat style

For information on creating and modifying styles, click the Contents tab in ArcGIS Desktop Help and navigate to Professional Library > Mapping and Visualization > Symbols and styles.

Having invested time in symbolizing a layer, you may want to save it as a layer file (a file with a .lyr extension). As you saw in chapter 4, layer files store symbology information for a spatial dataset.

Exercise 5c

In this exercise, you'll use symbols from the Conservation style to create a map that shows where elephants, giraffes, and zebras are found. Then you'll save the symbolized layer as a layer file.

1   Start ArcMap. In the ArcMap—Getting Started dialog box, under the Existing Maps section, click Browse for more. (If ArcMap is already running, click the File menu and click Open.) Navigate to **C:\ESRIPress\GTKArcGIS\Chapter05**. Click **ex05c.mxd** and click Open.

The map opens in layout view. You see the three data frames you've already symbolized and a fourth called Wildlife.

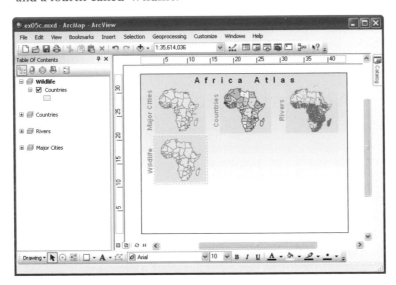

2   Click the View menu and click Data View.

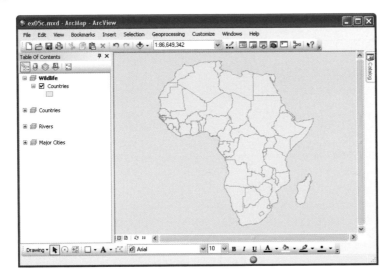

The active data frame contains a layer of countries. You'll add a layer of animal locations. In chapter 4, you added layers to ArcMap by dragging them from ArcCatalog. When ArcCatalog isn't open, you can use the Add Data button in ArcMap.

3 On the Standard toolbar, click the Add Data button.

4 In the Add Data dialog box, navigate to **\GTKArcGIS\Chapter05\Data**. You may have to click navigate through Folder Connections or Home—Chapter 05 to get to the Data folder. Click on **Animals.shp**. Make sure that your dialog box matches the following graphic, then click Add.

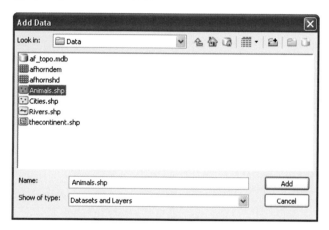

A point layer called Animals is added to the map. It displays with the default point symbol in a randomly chosen color.

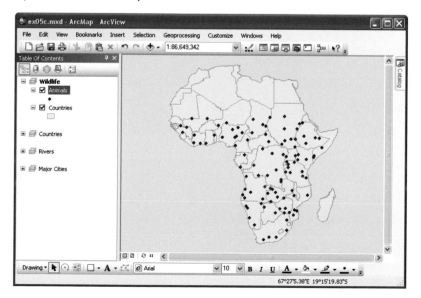

Each point in the Animals layer represents a significant population of elephants, giraffes, or zebras in the area around the point. Because the points have the same symbol, you can't tell which animal is which. You'll first symbolize the layer using an attribute of animal names, then you'll apply symbols for each type of animal.

5 Double-click the Animals layer. On the Layer Properties dialog box, click the Symbology tab. In the Show box, click Categories. Make sure the Unique Values option is highlighted.

The Value Field is already set to ANIMALNAME. Disregard the color scheme. You won't use it when you pick the animal symbols.

6 Click Add All Values. Values for elephants, giraffes, and zebras are added to the dialog box. In the Symbol column, click the check box next to <all other values> to uncheck it. Make sure your dialog box matches the following graphic, then click OK.

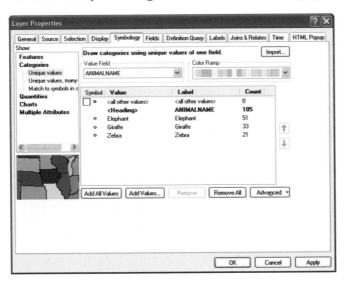

In the map, each kind of animal has its own color.

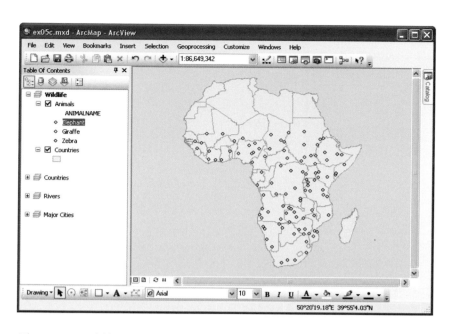

The map would be easier to read if the point symbols looked like the animals they represent.

7 In the table of contents, click the point symbol next to Elephant. The Symbol Selector opens. If you scroll through the list of symbols, you'll find there's nothing that looks like an elephant.

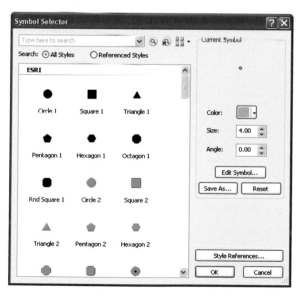

8   Click Style References to display the list of styles.

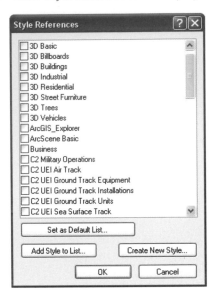

When a style is loaded, it has a check mark beside it. By default, only your personal style and the ESRI style are loaded. (Your personal style is empty unless you've added symbols to it.)

9   Click the check box beside Conservation style, then click OK to load it.

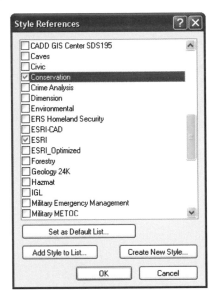

Symbols from the Conservation style are added below to the Symbol Selector. You can load as many styles in a map document as you want. The more you add, however, the longer it will take to scroll through the Symbol Selector. Search for a symbol if necessary.

10 Scroll down and click the Elephant symbol. In the Options frame, change the Size to 14 points. Click the Color square and click Delft Blue on the color palette. (The color palette now shows additional colors that belong to the Conservation style.) Click OK to close the Symbol Selector dialog box.

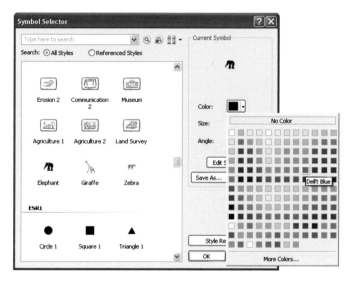

The elephant symbols are added to the map.

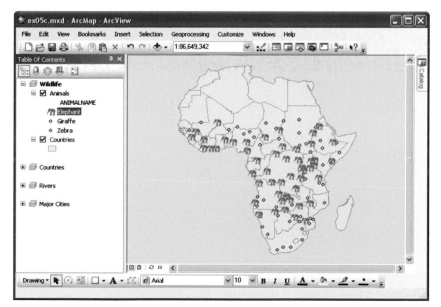

11   In the table of contents, click the point symbol next to Giraffe. In the Symbol Selector dialog box, type **giraffe** into the search box and click the search button. Click the Giraffe symbol.

12   In the Options frame, change the Size to 22. Click the Color square and click Raw Umber on the color palette. Make sure your dialog box matches the following graphic, then click OK in the Symbol Selector to add the giraffe symbols to the map.

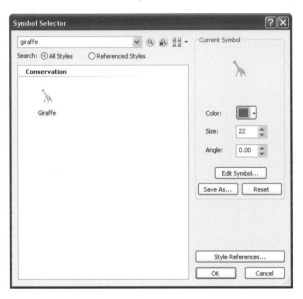

13   In the table of contents, click the Zebra symbol. In the Symbol Selector, search using the term **zebra** and click the Zebra symbol. You'll leave its color black and its size 18 points. Click OK to close the Symbol Selector dialog box.

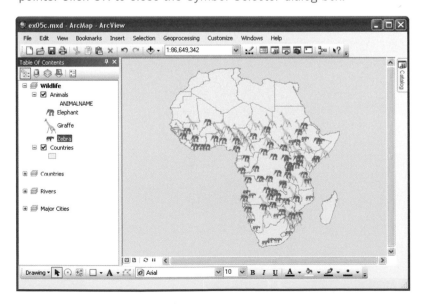

To use this Animals layer in another map—or send it to a colleague—without having to re-create the symbology, you can save it as a layer file.

14 In the table of contents, right-click the Animals layer and click Save As Layer File.

15 In the Save Layer dialog box, navigate to **\GTKArcGIS\Chapter05\MyData**. Accept the default file name of Animals.lyr. Make sure that your dialog box matches the following graphic, then click Save.

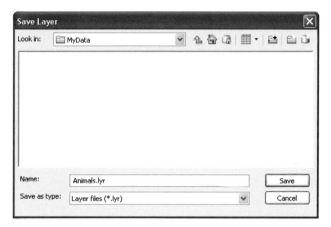

The Animals layer file can now be added to any map. Because it references the Animals.shp shapefile, the shapefile must also be accessible on disk or across a network. If you send the layer file to someone else, you must also send the data source it references.

You'll check your progress on the poster.

16 Click the View menu and click Layout View.

The poster is almost half finished. In the next exercise, you'll symbolize a raster dataset of elevation.

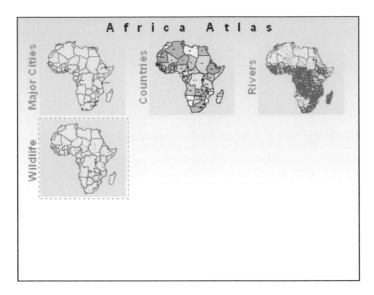

17 If you want to save your work, click the File menu and click Save As. Navigate to **\GTKArcGIS\Chapter05\MyData**. Rename the file **my_ex05c.mxd** and click Save.

18 If you are continuing with the next exercise, leave ArcMap open. Otherwise, exit the application. Click No if prompted to save your changes.

Symbolizing rasters

In chapter 1, you learned that a raster is a matrix of identically sized square cells. Each cell in a raster stores a value, usually a quantity of something such as elevation, rainfall, or temperature that has been measured at the location represented by the cell.

When you symbolize a raster, you assign colors to cell values or ranges of cell values. Raster values usually lie on a continuous scale, such as the scale of integers or real numbers, and are symbolized by color ramps.

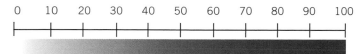

The shades of red in this color ramp become darker as the values increase from 0 to 100.

Rasters can represent many kinds of data and be symbolized with different color ramps.

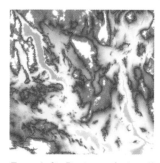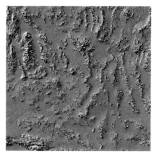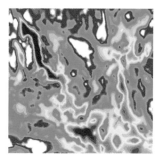

From left: Rasters of elevation, hillshade, and temperature for the same area. A hillshade raster shows an elevation surface in relief. It can be mathematically derived from an elevation raster by setting a certain angle and altitude of a light source.

Satellite images and aerial photographs are rasters in which the cell values are measurements of reflected light. Scanned maps are rasters in which the cell values are measurements made by the scanning device.

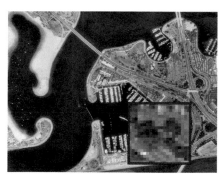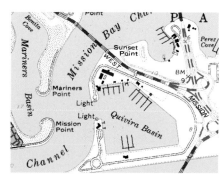

An air photo (left) and scanned map (right) of the area around San Diego's Mission Bay. If you zoom in close enough on any raster, its cell structure will be revealed, as shown by the inset in the red box on the air photo.

5
6
7

You can display raster data in ArcGIS Desktop, but to create it, you need one of the ArcGIS extensions for working with raster data—ArcGIS Spatial Analyst, ArcGIS 3D Analyst, or ArcGIS Geostatistical Analyst. These extensions are included on the software CD that comes with this book. For information about them, click the Contents tab in ArcGIS Desktop Help and navigate to Extensions. Raster data in many formats is also available on the Internet and from commercial vendors.

Exercise 5d

The next map in your Africa Atlas poster will be a topographical map (one that shows elevations or landforms). The map will include both vector and raster datasets. Because the rasters take up a lot of disk space, your study area will be confined to the Greater Horn of Africa, which includes Somalia, Ethiopia, Eritrea, Djibouti, and parts of other countries.

1   Start ArcMap. In the ArcMap—Getting Started dialog box, under the Existing Maps section, click Browse for more. (If ArcMap is already running, click the File menu and click Open.) Navigate to **C:\ESRIPress\GTKArcGIS\Chapter05**. Click **ex05d.mxd** and click Open.

The map opens in layout view. You see the four data frames you've already symbolized and a fifth called Topography.

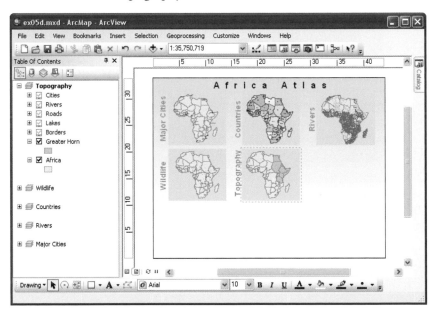

2   Click the View menu and click Data View.

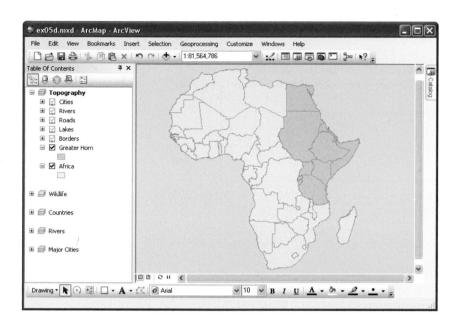

The active data frame contains seven layers. Countries in the study area are symbolized in orange. The Cities, Rivers, Roads, Lakes, and Borders layers are scale-dependent. You'll see them when you zoom in.

Now you'll add elevation and hillshade raster layers to the map.

3   On the Standard toolbar, click the Add Data button.

4   In the Add Data dialog box, navigate to **\GTKArcGIS\Chapter05\Data**. Click **afhorndem**. Hold down the Control (Ctrl) key and click **afhornshd**. Make sure your dialog box matches the following graphic, then click Add.

A message prompts you to create pyramids for afhorndem.

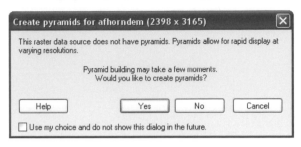

5 Click Yes to create pyramids for afhorndem.

When the process is finished, you're prompted to build pyramids for the other raster layer.

6 Click Yes to create pyramids for afhornshd.

Building pyramids

Pyramids are versions of a raster dataset, varying from coarse to fine resolution, that are used to improve the drawing speed of raster layers as you zoom in or out. Coarse-resolution versions are used when you are zoomed at or near the full extent; finer-resolution versions are used as you zoom in. The coarseness of the resolution corresponds to the amount of detail you would expect to see at a given scale. You have to build pyramids only once for a raster dataset. They are stored with the data as a file with the extension .rrd. For more information, click the Help button on the Create pyramids dialog box or click the Contents tab in ArcGIS Desktop Help and navigate to *Professional Library > Geoprocessing > Geoprocessing tool reference > Data Management toolbox > Raster toolset > Raster Dataset toolset > Raster properties toolset > Build pyramids*.

The raster layers are added to the bottom of the table of contents. You'll rename them.

7 In the table of contents, double-click the afhornshd layer. In the Layer Properties dialog box, click the General tab. In the Layer Name text box, the name afhornshd should be highlighted. Type **Hillshade** in its place.

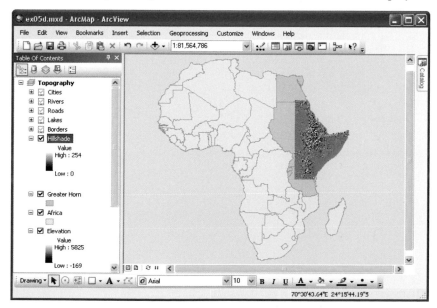

8 Click OK.

9 In the table of contents, double-click the afhorndem layer. In the Layer Properties dialog box, click the General tab if necessary. Rename the layer **Elevation** and click OK.

To see the rasters, you need to change their position in the table of contents.

10 In the table of contents, click the Hillshade layer. Drag it above the Greater Horn layer. (As you drag, the current position of the layer is shown by a horizontal black bar.) Release the mouse button to drop the layer.

The Hillshade layer is repositioned in the table of contents and displays on the map.

11 In the table of contents, drag the Elevation layer above the Hillshade layer.

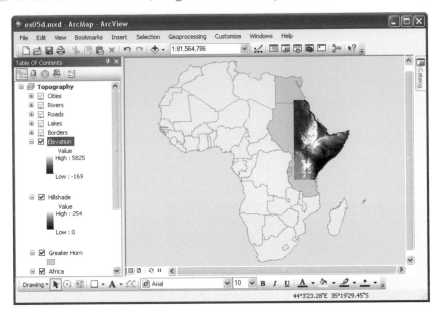

By default, rasters are drawn in shades of gray. You'll choose a different color ramp for the Elevation layer.

12 In the table of contents, right-click the Elevation layer and click Zoom to Layer. The map zooms in.

13 In the table of contents, double-click the Elevation layer. In the Layer Properties dialog box, click the Symbology tab.

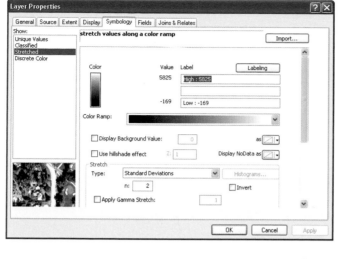

There are four symbology methods in the Show box. In the Stretched method (the default), ArcMap makes subtle transitions along the selected color ramp, but doesn't give you precise information about which data values are associated with which shades of color. In the Classified method, color transitions are less subtle, but you can see exactly which value ranges correspond to which shades. In the Unique Values method, a different color is assigned to every data value in the raster. This method is available as long as the raster has no more than 2,048 unique data values. In the Discrete Color method, a color is assigned to each unique value until it reaches the maximum number of colors you choose. It is useful when a very large number of unique values are present.

14 In the Color Ramp drop-down list, right-click the colors (not the drop-down arrow). On the context menu, click Graphic View to uncheck it. Click the Color Ramp drop-down arrow. Scroll down to the Elevation #1 color ramp and click it. Make sure your dialog box matches the following graphic, then click OK to close the Layer Properties.

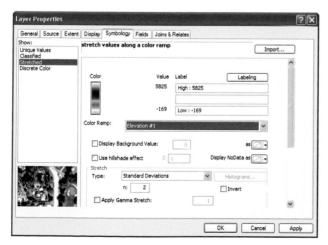

The Elevation #1 color ramp is applied to the layer. The colors grade from light blues at lower elevations through yellows, greens, oranges, and browns to gray and white at higher elevations.

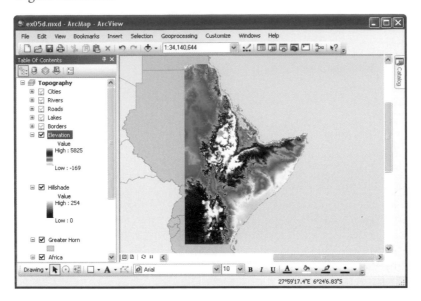

To see surface relief you will make the Elevation layer partially transparent so that the Hillshade layer shows through.

15 Double-click the Elevation layer. In the Layer Properties dialog box, click the Display tab. In the Transparency text box, highlight the default value of 0 and type **70**. The Elevation layer will be 70 percent transparent. Make sure your dialog box matches the following graphic, then click OK.

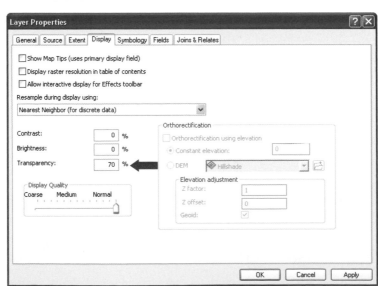

The Hillshade layer gives a realistic impression of the landform.

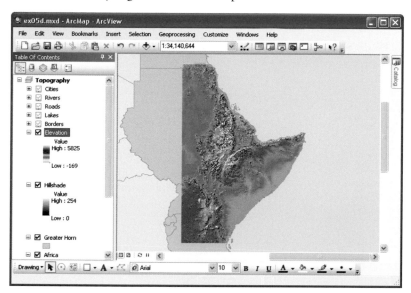

You'll zoom in to see the vector layers displayed on top of the raster layers.

16 Click the Bookmarks menu and click Closeup. The map zooms in and the scale-dependent layers display.

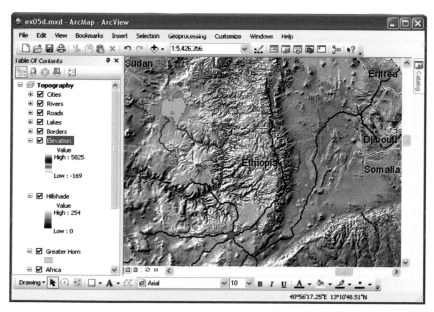

17 In the table of contents, right-click the Elevation layer and click Zoom To Layer. The display zooms out to the Greater Horn.

You'll examine your progress on the Africa Atlas poster.

18 Click the View menu and click Layout View.

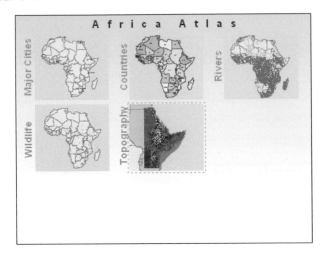

In the next chapter, you'll symbolize layers by quantitative attributes rather than categorical ones. You'll also learn how to classify data; that is, to adjust the value ranges to which symbols are applied. And you'll work with symbology methods that are used specifically with numeric data.

19 If you want to save your work, click the File menu and click Save As. Navigate to **\GTKArcGIS\Chapter05\MyData**. Rename the file **my_ex05d.mxd** and click Save.

20 If you are continuing to the next chapter, leave ArcMap open. Otherwise, exit the application. Click No when prompted to save your changes.

Chapter 6

Classifying features and rasters

Classifying features by standard methods
Classifying features manually
Mapping density
Using graduated and chart symbols

With a few exceptions, like dates, feature attributes that are not categorical are quantitative—counted, measured, or estimated amounts of something. The population of a country, the length of a river, and the estimated size of an oil deposit are all quantitative attributes. Quantitative attributes are always numeric.

When you symbolize quantities, you want to see where attribute values lie in relation to one another on a continuous scale. Unless you are working with a very small range of values, the values must be divided into groups to make the symbology manageable. Dividing values into groups—or classes—requires that you choose both the number of classes and a method to determine where one class ends and another begins. Making these two decisions is called classifying data.

Because quantitative values are on a number scale (whether of integers, real numbers, or a specialized scale such as degrees Fahrenheit), the symbology applied to them is usually also scaled. ArcGIS has four ways to apply scaled symbology: by graduated color, graduated symbol, proportional symbol, and dot density.

Graduated color symbology, the most common, displays features as shades in a range of colors that changes gradually. The range is called a color ramp. If you symbolized the countries of Africa by population, each country would be drawn in a different shade of a color, such as blue, according to its population. Countries with large populations, like Nigeria, would be dark blue, while countries with small populations, like Djibouti, would be light blue. Graduated color symbology is most effective on polygon layers because subtle color differences are easier to detect on large features.

Graduated symbols represent features using different marker sizes. Normally used with point layers—to indicate, for instance, the population of a city—graduated symbols can also be used with lines or polygons. In the case of polygons, markers are drawn inside the features.

Proportional symbols vary in size proportionally to the value symbolized. For example, the marker for a city of 10,000,000 would be ten times larger than the marker for a city of 1,000,000. Proportional symbols work best when the range of values for an attribute is not too wide.

Dot density, available for polygon layers only, represents quantities by a random pattern of dots. The greater the value, the more dots are displayed within the feature boundary. Like proportional symbols, dot density maps convey quantities precisely because there is a fixed relationship between the number of dots and the attribute symbolized (for example, one dot equals 10,000 people). They can be misleading, however, because the random distribution of dots may be different from the actual distribution of values.

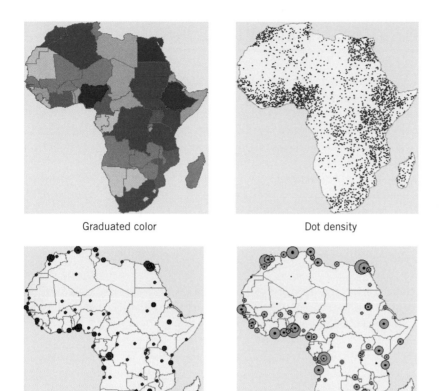

Graduated color

Dot density

Graduated symbol

Proportional symbol

Upper left, population by country: Darker shades of red indicate larger populations.

Upper right, relative population density: Dot clusters show densely populated areas.

Lower left, population for selected cities: Each city falls into one of four population groups, and each group is assigned a different symbol size.

Lower right, population for selected cities: The size of the symbol is proportional to each city's population.

Classifying features by standard methods

ArcGIS has seven classification methods: natural breaks (Jenks), equal interval, defined interval, quantile, standard deviation, geometric interval, and manual.

Natural breaks is the default method. Developed by the cartographer George Jenks, it creates classes according to clusters and gaps in the data.

Equal interval creates classes of equal value ranges. If the range of values is 1 to 100 and the number of classes is 4, this method will create classes from 1–25, 26–50, 51–75, and 76–100.

Defined interval resembles equal interval except that the interval determines the number of classes rather than the other way around. If the range of values is 1 to 100, and you choose an interval of 10, this method will create 10 classes: 1–10, 11–20, 21–30, and so on.

Quantile creates classes containing equal numbers of features. If you choose five classes for a layer with 100 features, this method will create class breaks so that 20 features fall into each class. The value range varies from class to class.

Standard deviation creates classes according to a specified number of standard deviations from the mean value.

Geometric intervals creates classes based on class intervals that have a geometric series. A geometric series is a pattern where a constant coefficient multiplies each value in the series. It produces a result that is visually appealing, cartographically comprehensive, and minimizes variance within classes.

With the manual method, you can set whatever class breaks you like.

Exercise 6a

In chapter 5, you created five maps for the Africa Atlas. You have four maps to complete. In this exercise, you'll make a population map and apply different classification methods and numbers of classes to it.

1   Start ArcMap. In the ArcMap—Getting Started dialog box, under the Existing Maps section, click Browse for more. (If ArcMap is already running, click the File menu and click Open.) Navigate to **C:\ESRIPress\GTKArcGIS\Chapter06**. Click **ex06a.mxd** and click Open.

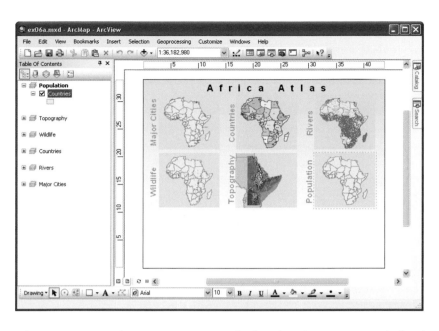

The map opens in layout view. You see the five data frames you symbolized in chapter 5. The sixth data frame in the atlas is called Population. It has a dashed line around it because it is active. It is at the top of the table of contents.

2   Click the View menu and click Data View.

3   In the table of contents, right-click the Countries layer in the Population data frame and click Open Attribute Table.

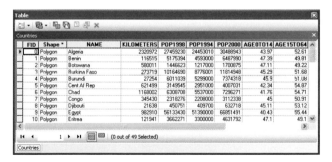

You'll use the POP2000 attribute because it has the most current data.

4   Right-click the POP2000 field name and click Sort Ascending. Scroll down through the table.

About half the countries have populations under ten million. Just a few have populations over forty million. Nigeria, with 161 million people, has more than double the population of Ethiopia.

5 Close the attribute table. In the table of contents, double-click the Countries layer name to open the Layer Properties dialog box. Click the Symbology tab.

6 In the Show box, click Quantities.

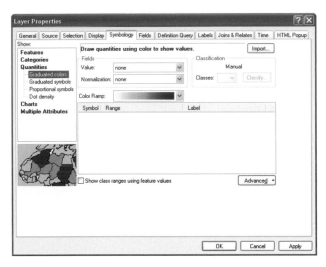

The Graduated colors option of the Quantities method is highlighted and the dialog box changes. The Value field drop-down list shows the available attributes—numeric fields only.

7 Click the Value drop-down arrow and click POP2000.

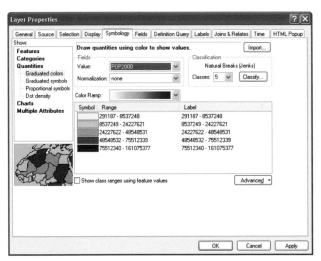

By default, the values in the POP2000 field are grouped into five classes and the classification method is Natural Breaks.

8 If your color ramp is different from the one in the previous graphic, click the Color Ramp drop-down list and click the Yellow to Dark Red ramp. (To find it by name, right-click the ramp and uncheck Graphic View.) Click OK to close the Layer Properties dialog box.

The symbology is applied to the map. The greater a country's population, the darker its color.

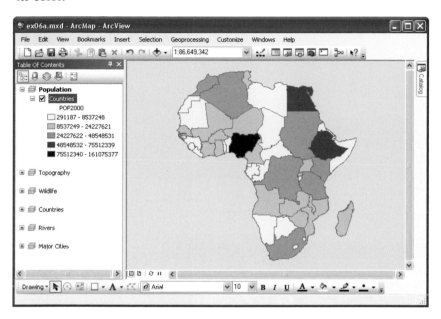

The number of features in each class is different. The class with the highest values has just one member: Nigeria. The next highest class has just two: Egypt and Ethiopia. The remaining 46 countries belong to the three lowest classes. This map shows that a few countries in Africa have much greater populations than the rest. Use the Identify tool if you want to look at the information of other countries.

9 In the table of contents, double-click the Countries layer to open the Layer Properties dialog box. Click the Symbology tab if necessary.

10 Click the Classes drop-down arrow and click 3. The number of classes and their value ranges are adjusted. Click Apply and move the Layer Properties dialog box away from the map display.

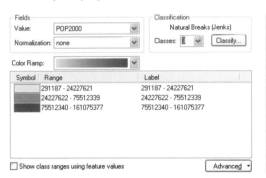

The map reflects the new classification. The highest population class still contains just a single country, but many countries that belonged to separate classes before are now grouped together. This map focuses attention on Nigeria as the most populous country on the continent.

11 On the Symbology tab, click Classify.

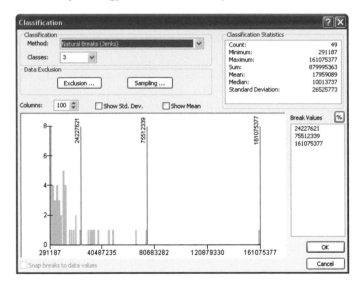

The Classification dialog box displays the current classification method and the number of classes. You can also see statistics for the POP2000 field. The large window is a histogram, or frequency distribution chart.

Understanding the histogram

The x-axis (horizontal) shows the range of values in the field. The y-axis (vertical) is a count of features. Vertical blue lines are class breaks (also shown in the Break Values box).

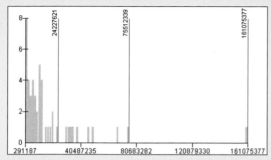

The gray columns represent percentages of the value range. The default number of gray columns is one hundred. In this case, the lowest value on the x-axis is 291,187 and the highest is 161,075,377. The value range is therefore 160,784,190, of which 1 percent (the amount represented by a single gray column) is 1,607,842.

The first gray column has a y-value of 8. This means that eight countries have populations under 1,899,029 (the start value of 291,187 plus 1,607,842). Another way to put it is that eight countries fall within the bottom 1 percent of the value range.

The second column has a y-value of 4. This means that four countries have populations over 1,899,029 but under 3,506,871 (1,899,029 plus 1,607,842). Cumulatively, twelve countries fall within the bottom 2 percent of the range.

The third column has a y-value of 3. Three countries have populations over 3,506,871 but under 5,114,713. Cumulatively, fifteen countries (of forty-nine total) fall within the bottom 3 percent of the range. The data is heavily skewed toward the low end.

A gray column isn't drawn if no features fall within the value range it would represent. The number of columns can be set to anything from ten to one hundred. With ten columns, for instance, each would represent 10 percent of the value range.

12 In the Columns box, highlight the default value of 100 and type **25**.

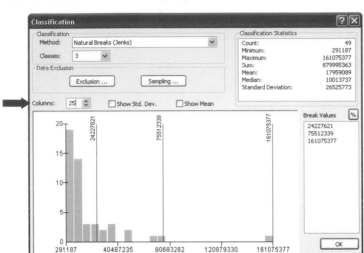

Now each gray column represents one twenty-fifth (or 4 percent) of the value range.

The histogram shows you where class breaks fall in relation to the data. The classification method is set to Natural Breaks (Jenks) where breaks reflect clusters in the data. For example, the population of Nigeria is more than twice that of any other country and is placed in a class by itself.

13 In the Classification frame, click the Method drop-down arrow and click Equal Interval. The class breaks are adjusted.

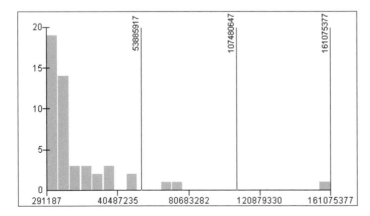

This method places breaks so that each class has the same range of values. In this case, almost all the values lie in the first class. Unlike natural breaks, it's possible with equal interval to have classes with no values in them. (You would see this if you increased the number of classes to four.)

14 Click the Method drop-down arrow and click Quantile.

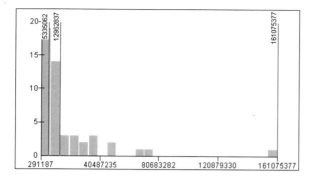

The quantile method places breaks so that each class contains an approximately equal count of values. (In other words, each class includes about the same number of features.)

15 Click OK in the Classification dialog box. The current classification settings are displayed on the Symbology tab. Click Apply and move the Layer Properties dialog box away from the map display.

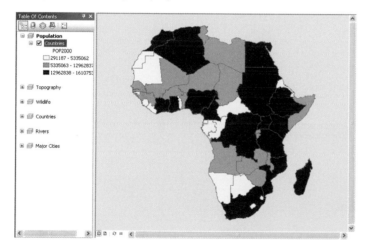

On the map, each population class includes a roughly equal number of countries. This map suggests—falsely—that countries with similar colors have similar population values. In fact, the value ranges for the three classes are very different.

For your final population map, you'll return to the natural breaks method.

16 On the Symbology tab, click Classify.

17 In the Classification dialog box, click the Method drop-down arrow and click Natural Breaks (Jenks). Click the Classes drop-down arrow and click 7. Make sure your dialog box matches the following graphic, then click OK.

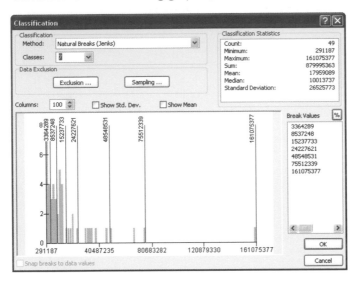

Before applying the changes to the map, you'll make the class labels easier to read by putting commas in.

18 On the lower portion of the Symbology tab, click the Label column heading and click Format Labels. The Number Format dialog box opens.

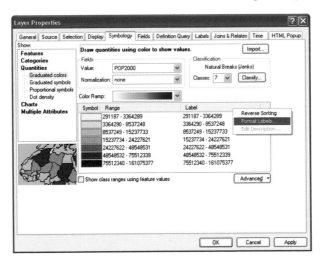

The Number Format dialog box lets you adjust the formatting of numeric labels in many ways—you can control the number of decimal places displayed, insert commas to separate thousands, and change the alignment. You can also display numbers with special notation to indicate currency, percentages, and so on.

19 In the Number Format dialog box, check the Show thousands separators check box. Click OK to close the dialog box.

20 At the bottom of the Symbology tab, click the check box to Show class ranges using feature values. Every class start and end now corresponds to a feature value. Make sure your Layer Properties dialog box matches the following graphic, then click OK.

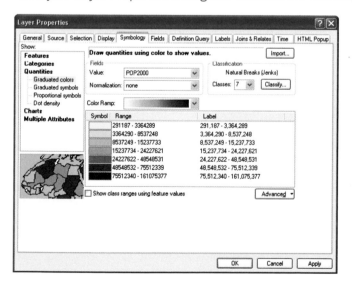

21 In the table of contents, scroll to the right, if necessary, to see the class ranges in their entirety.

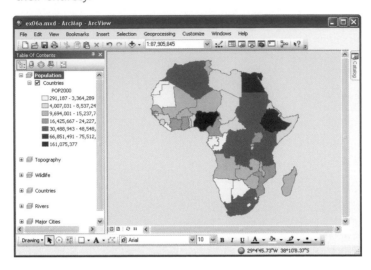

Countries are grouped with other countries that have similar populations. Nigeria remains in a class by itself. In the table of contents, the class breaks now indicate where there are gaps in the data. You can see, for instance, that there are no countries with populations between 49 and 66 million or between 76 and 161 million.

You'll check your progress on the Africa Atlas poster.

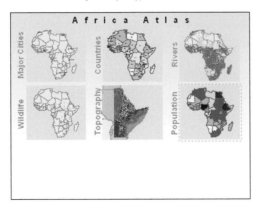

22 Click the View menu and click Layout View.

23 If you want to save your work, click the File menu and click Save As. Navigate to **\GTKArcGIS\Chapter06\MyData**. Rename the file **my_ex06a.mxd** and click Save.

24 If you are continuing with the next exercise, leave ArcMap open. Otherwise, exit the application. Click No if prompted to save your changes.

Classifying features manually

Manual classification is used to reveal significant groupings in data that standard classification methods miss. For example, a market researcher might want to distinguish census tracts in which the average household income is more than $100,000 a year or the 18 to 49 age group makes up a certain percentage of the population. To symbolize these values specifically, you would need to set class breaks manually.

Exercise 6b

In this exercise, you won't add a new map to the atlas. Instead, you'll make a refinement to the Topography map from chapter 5. You'll manually create a class to represent elevations below sea level.

1 Start ArcMap. In the ArcMap—Getting Started dialog box, under the Existing Maps section, click Browse for more. (If ArcMap is already running, click the File menu and click Open.) Navigate to **C:\ESRIPress\GTKArcGIS\Chapter06**. Click **ex06b.mxd** and click Open.

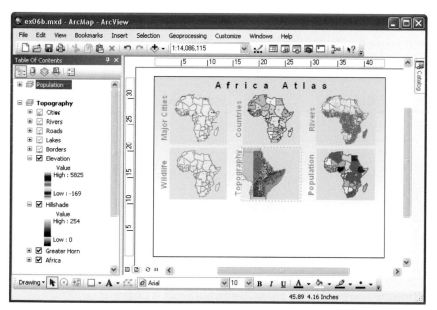

The map opens in layout view. You see the six data frames you've already symbolized.

2 Click the View menu and click Data View.

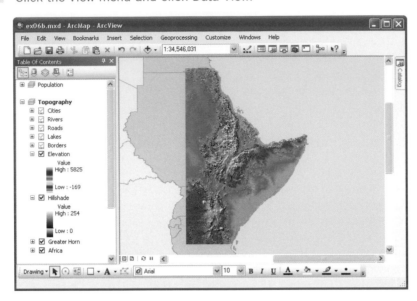

The Topography data frame is active. In the table of contents, you see that the elevation values range from a high of 5,825 to a low of -169 (in meters). You'll change the classification so that elevation values below sea level belong to their own class.

3 Click the Bookmarks menu and click Elevation Area.

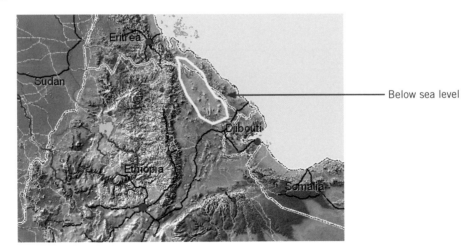

The map zooms in and the vector layers display. (The vector layers display at scales of 1:10,000,000 or larger. If you don't see them, zoom in until you do.)

The area below sea level, outlined in yellow in the book (but not on your screen), lies along the northeastern border of Ethiopia and Eritrea.

4   In the table of contents, double-click the Elevation layer. In the Layer Properties dialog box, click the Symbology tab.

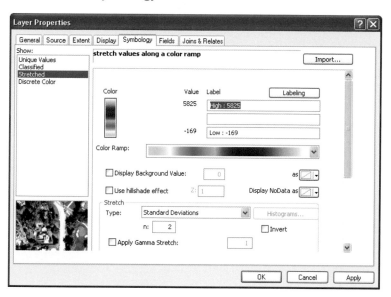

To classify the values, you'll change the method to Classified.

5   In the Show box, click Classified.

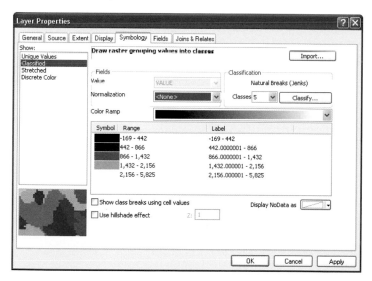

By default, there are five classes and the classification method is Natural Breaks.

6   Click the Classes drop-down arrow and click 10. Right-click the color ramp and uncheck Graphic View. Click the drop-down arrow, then scroll down and set the color ramp to Elevation #2.

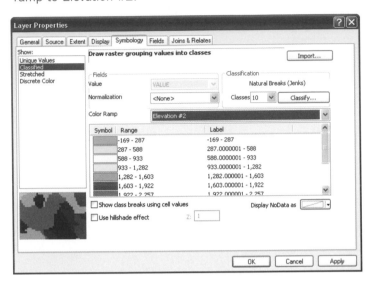

The first class has a range of -169 to 287. You'll reset this break so that the class values range from -169 to 0.

7   Click Classify.

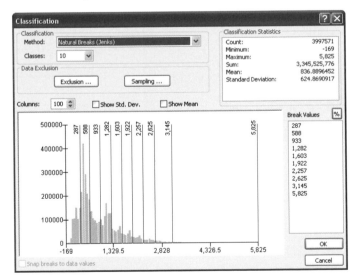

The histogram shows the distribution of values. The extreme high values belong to Mount Kilimanjaro.

8 In the Break Values box, click the first value, 287, to highlight it. The blue line in the histogram corresponding to that value turns red. Type **0** and then click the empty space at the bottom of the Break Values box.

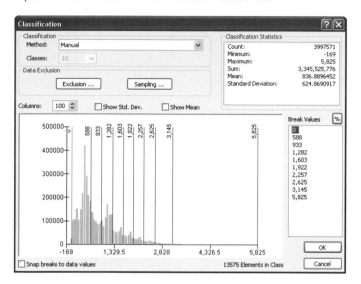

In the histogram, the class break moves to the left and is relabeled 0. In the Method drop-down list, the method automatically changes to Manual.

Manual classification

Replacing values in the Break Values box is one way to adjust classes manually. You can also drag the blue class break lines in the histogram to new positions, right-click a class break to delete it, or right-click any empty part of the histogram to insert a new break.

To make sure each class begins and ends with a value that exists in the attribute table, you can check Snap breaks to data values at the bottom of the Classification dialog box. This usually results in gaps between classes, but more accurately describes the value distribution. For more information, click the Contents tab in ArcGIS Desktop Help and navigate to *Professional Library > Extensions > Geostatistical Analyst > Visualizing and managing geostatistical layers > Classifying data > …by manually altering the class breaks.*

9   Click OK in the Classification dialog box. The Symbology tab is updated.

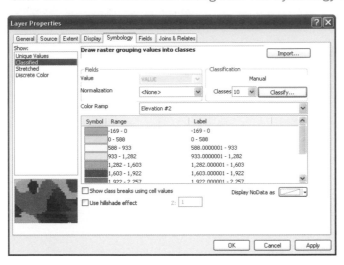

To make the areas below sea level show up better on the map, you'll symbolize them with a brighter color.

10   In the Symbol column, double-click the light blue symbol assigned to the lowest elevation class (-169–0). In the color palette, click Mars Red.

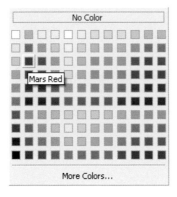

Finally, you'll format the labels. You don't need to show all those decimal places.

11 Click the Label column heading and click Format Labels. In the Rounding frame of the Number Format dialog box, click the Number of decimal places option. Highlight the default value of 10 and type **0**. Make sure the Show thousands separators box is checked.

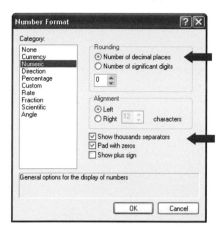

12 Click OK in the Number Format dialog box. Make sure your Layer Properties dialog box matches the following graphic, then click OK.

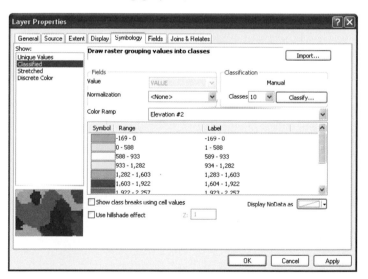

13 Click the Bookmarks menu and click Below Sea Level. The area below sea level displays in Mars Red. (It isn't as bright as you might have expected, because the layer is 70 percent transparent.)

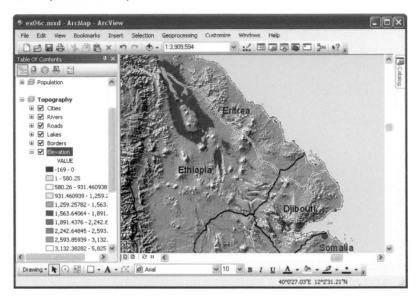

14 In the table of contents, right-click the Elevation layer and click Zoom To Layer.

The view zooms back to the scale that will be displayed on your poster. The areas below sea level are still faintly visible.

You'll look at your Africa Atlas poster.

15 Click the View menu and click Layout View.

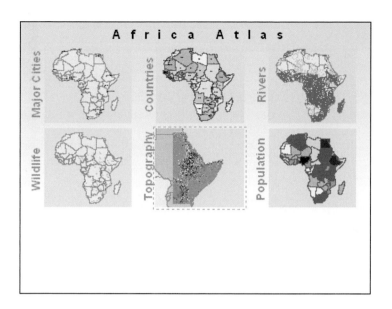

The poster is two-thirds finished. In the next exercise, you'll symbolize population density.

16 If you want to save your work, click the File menu and click Save As. Navigate to **\GTKArcGIS\Chapter06\MyData**. Rename the file **my_ex06b.mxd** and click Save.

17 If you are continuing with the next exercise, leave ArcMap open. Otherwise, exit the application. Click No if prompted to save your changes.

Mapping density

Nigeria has more than 160 million people. Rwanda has about 10 million. But which of the two is more densely populated? To answer that question, you would need to know the number of people per square unit of area. As long as you have both a population attribute and an area attribute, this is a straightforward operation—you simply divide population by area. Dividing one attribute by another to find the ratio between them is called normalization. It is commonly used to calculate density, but has other uses as well. For example, normalizing population by income gives income per capita.

Exercise 6c

In this exercise, you'll normalize population by area to make a population density map. You'll symbolize the map with graduated colors and by dot density.

1 Start ArcMap. In the ArcMap—Getting Started dialog box, under the Existing Maps section, click Browse for more. (If ArcMap is already running, click the File menu and click Open.) Navigate to **C:\ESRIPress\GTKArcGIS\Chapter06**. Click **ex06c.mxd** and click Open.

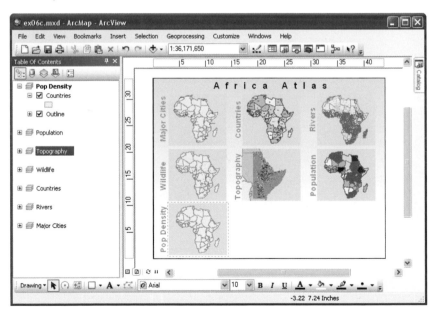

The seventh data frame in the Africa Atlas is called Pop Density.

2 Click the View menu and click Data View.

In the table of contents, the Pop Density data frame is active.

3 In the table of contents, right-click the Countries layer in the Pop Density data frame and click Open Attribute Table.

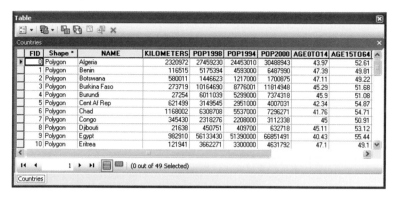

The KILOMETERS attribute contains the size of each country in square kilometers. You'll divide the values in the POP2000 field by these values to obtain the population density of each country.

4 Close the table. Double-click the Countries layer to open the Layer Properties dialog box. Click the Symbology tab.

5 In the Show box, click Quantities. The Graduated colors option is highlighted. In the Fields frame, click the Value drop-down arrow and click POP2000. Click the Normalization drop-down arrow and click KILOMETERS.

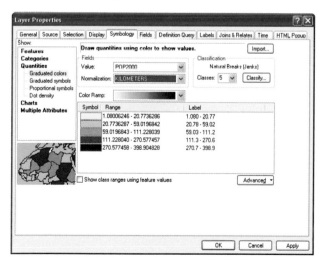

The values in the Range column now express population per square kilometer. In the Labels column, values are rounded, the default for normalized data.

6 If necessary, set the color ramp to Yellow to Dark Red. Click OK to close the Layer Properties dialog box.

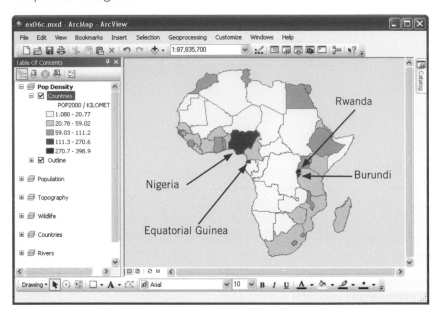

The class with the greatest population density again has a single member, but this time it's tiny Rwanda. The next highest class contains two other small countries, Burundi and Equatorial Guinea, along with Nigeria. Several countries with large populations turn out to have low population densities. (The labels that are shown in the book are not on your screen.)

Graduated color maps are one way to represent density. Another way is with a dot density map.

7 Double-click the Countries layer to open the Layer Properties dialog box. The Symbology tab is active.

8 In the Show box, click the Dot density option under Quantities.

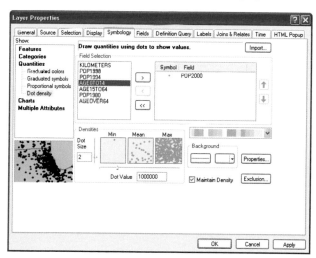

On the left side of the Field Selection frame, numeric fields in the attribute table are listed in a box. You are symbolizing the POP2000 field.

9 Click POP2000 to highlight it. Click the right arrow symbol (>). The field is added to the box on the right and a dot symbol is assigned to it. (The color of your symbol may be different than the one shown in the graphic.)

Next, you will set the symbol color for the dots. You will also turn the country outlines off to make the map easier to read.

10 In the Symbol column, right-click the dot symbol. On the color palette, click Gray 60%.

11 In the Background frame, click the Line button.

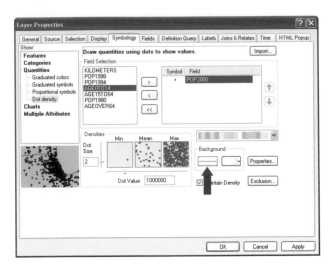

The Symbol Selector dialog box opens.

12 Click the Color square. On the color palette, click No Color. Click OK to close the Symbol Selector dialog box. This turns off the country outlines.

13 In the Densities frame of the Symbology tab, highlight the dot value and type **750000**. The map will display one dot for every 750,000 people. Make sure the dialog box matches the following graphic, then click OK.

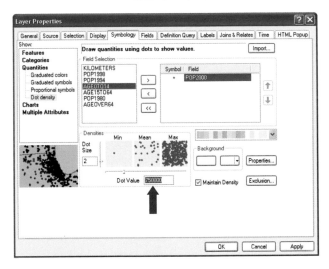

The map shows, as you would expect, that the coastal countries of Africa are more densely populated than the interior.

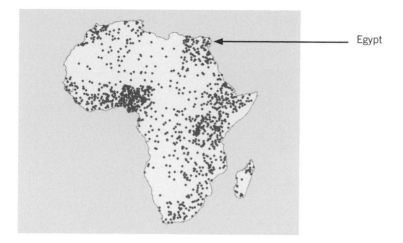

Egypt

Because the dot placement is random within each country, the arrangement of dots on your map won't match the graphic exactly. Dot density maps reveal meaningful overall patterns, but the random distribution of dots can be misleading if you focus on small areas. For example, the dots in Egypt are distributed throughout the country, when, in fact, almost all of Egypt's population lives along the Nile River.

Controlling dot placement

Clicking Properties on the Symbology tab opens the Dot Density Symbol Properties dialog box. By checking the Use Masking check box, you can specify either that dots be placed only within the mask area or that they be excluded from it. For example, you might use a lakes layer as a mask for a population density map to ensure that no dots were placed within lakes. Conversely, you might use a buffer layer to ensure that all dots were placed within the area of the buffer. (You will learn about buffers in chapter 12.)

You'll check your progress on the Africa Atlas.

14 Click the View menu and click Layout View.

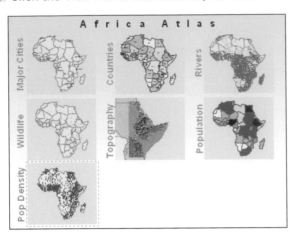

15 If you want to save your work, click the File menu and click Save As. Navigate to **\GTKArcGIS\Chapter06\MyData**. Rename the file **my_ex06c.mxd** and click Save.

16 If you are continuing with the next exercise, leave ArcMap open. Otherwise, exit the application. Click No if prompted to save your changes.

Using graduated and chart symbols

Graduated symbols use diffcrent marker sizes to represent features.

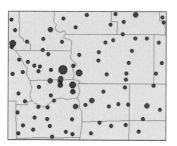 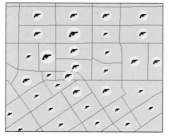 

Left: A point layer of towns symbolized by population. Center: A polygon layer of neighborhoods symbolized by number of crimes. Right: A street layer symbolized by number of lanes per street.

Quantitative attributes are usually symbolized by one of the four Quantities options, but they can also be symbolized as Charts. This method allows you to symbolize several attributes at once by drawing pie, bar, or stacked bar charts on each feature in a layer.

Pie charts symbolize the percentages that various attributes contribute to a total. They might be used to show how much of a country's gross domestic product comes from agricultural production, industry, and services. They might show the percentages of a country's population that are 0 to 14 years old, 15 to 64, and 65 or older.

Bar charts compare attributes in cases where the values do not make up a whole. They might be used to compare a country's 1990 population to its 2000 population or income to spending.

Stacked bar charts represent the cumulative effect of attributes. For example, if you wanted to display the total number of personal communication devices in a country, you could symbolize the values for three different attributes—main-line telephones, cellular telephones, and computers with Internet access—as a stacked bar chart.

Exercise 6d

In this exercise you'll create the last two maps for the Africa Atlas. The first will use graduated symbols to show the size of diamond mines. The second will use pie charts to show the makeup of each country's electrical production.

1   Start ArcMap. In the ArcMap—Getting Started dialog box, under the Existing Maps section, click Browse for more. (If ArcMap is already running, click the File menu and click Open.) Navigate to **C:\ESRIPress\GTKArcGIS\Chapter06**. Click **ex06d.mxd** and click Open.

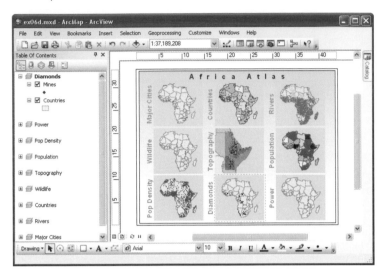

The last two data frames of the atlas will show the location of diamond mines and the sources of electric power.

2   Click the View menu and click Data View.

The points represent the locations of forty-eight diamond mines. Because all the point symbols are the same, you can't tell anything about the size of each mine.

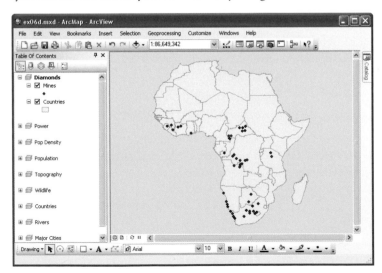

3   In the table of contents, right-click the Mines layer and click Open Attribute Table.

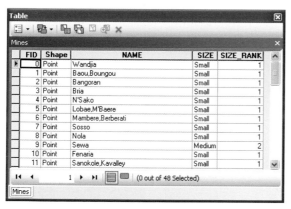

The layer attributes include the name of the mine, its size (small, medium, or large), and a SIZE_RANK field that translates the size values into numbers.

4   Close the table. In the table of contents, double-click the Mines layer to open the Layer Properties dialog box. Click the Symbology tab.

5   In the Show box, click Quantities, then click Graduated symbols.

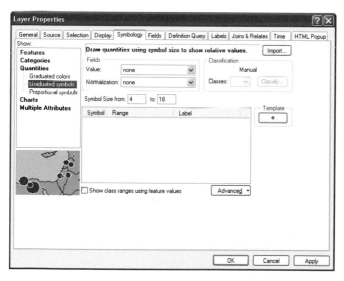

6 Click the Value drop-down arrow and click SIZE_RANK (the only numeric field in the table).

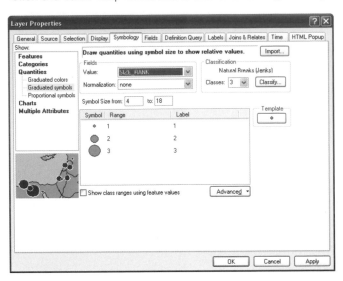

In the symbol column, three symbols appear. You'll change them to a symbol that looks more like a diamond.

7 Click the Template button to open the Symbol Selector.

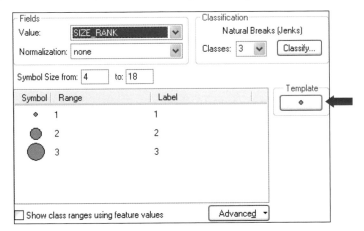

8 In the Symbol Selector, click the Square 2 symbol. Highlight the Angle value of 0.00 and type **45**. Click the Color square. On the color palette, click Electron Gold. Click OK to close the Symbol Selector dialog box.

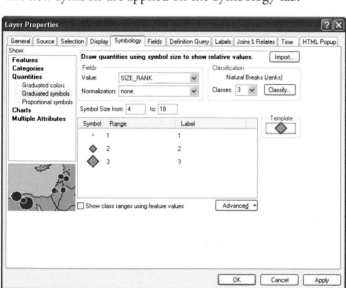

The new symbols are applied on the Symbology tab.

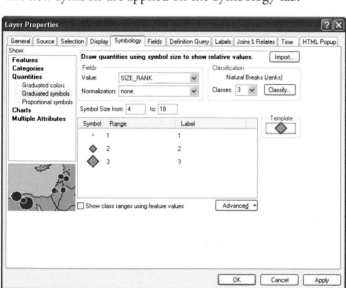

You'll change the symbol sizes slightly.

9   In the Symbol Size from boxes, highlight 4 and change it to **5**. Highlight 18 and change it to **16**.

You'll also replace the numeric labels with descriptions.

10 In the Label column, click "1." An input box opens with the label highlighted in it. Type **Small** as the new label.

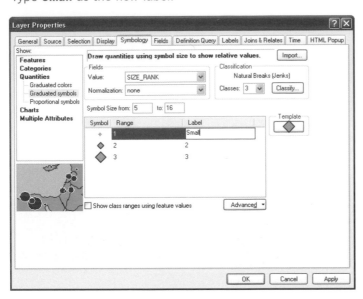

11 Press Enter to advance to the label "2" and replace it with **Medium**. Press Enter and replace the label "3" with **Large**. Click the empty white space to close the last input box. Make sure the dialog box matches the following graphic, then click OK.

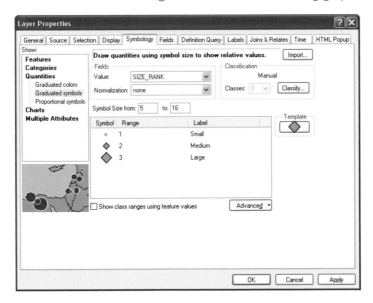

The map now shows the locations and sizes of diamond mines across the continent.

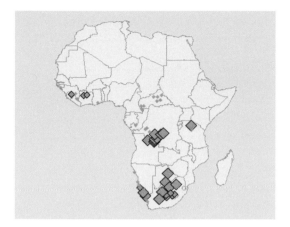

Now you'll use chart symbols to show the sources of electrical power in each country.

12 In the table of contents, click the minus sign by the Diamonds data frame. Right-click the Power data frame and click Activate. Click the plus sign by the Power data frame.

13 Right-click the Countries layer and click Open Attribute Table. Scroll all the way to the right.

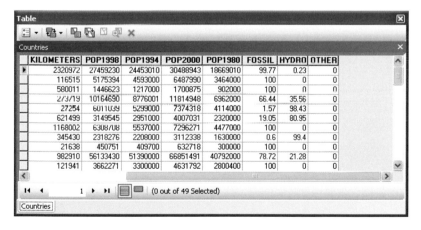

KILOMETERS	POP1998	POP1994	POP2000	POP1980	FOSSIL	HYDRO	OTHER
2320972	27459230	24453010	30488943	18669010	99.77	0.23	0
116515	5175394	4593000	6487990	3464000	100	0	0
580011	1446623	1217000	1700875	902000	100	0	0
273719	10164690	8776001	11814948	6962000	66.44	35.56	0
27254	6011039	5299000	7374318	4114000	1.57	98.43	0
621499	3149545	2951000	4007031	2320000	19.05	80.95	0
1168002	6308708	5537000	7296271	4477000	100	0	0
345430	2318276	2208000	3112338	1630000	0.6	99.4	0
21638	450751	409700	632718	300000	100	0	0
982910	56133430	51390000	66851491	40792000	78.72	21.28	0
121941	3662271	3300000	4631792	2800400	100	0	0

(0 out of 49 Selected)

Countries

The last three fields show what percentage of each country's electrical production comes from fossil fuels, hydroelectricity, and other means. You'll use pie charts to represent this data.

14 Close the table. Double-click the Countries layer to open the Layer Properties dialog box. Click the Symbology tab.

15 In the Show box, click Charts. Make sure the Pie option is highlighted. The dialog box changes to show the choices for pie charts. The Field Selection box shows all numeric attributes in the table.

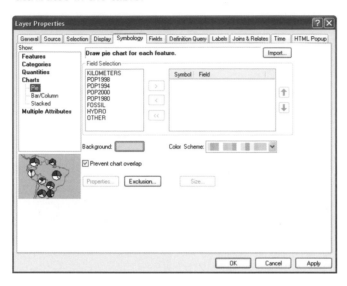

It doesn't matter if your Background and Color Scheme settings are different from those in the graphic. You will change these settings in a moment.

16 In the Field Selection box, click FOSSIL. Hold down the Control (Ctrl) key and click HYDRO and OTHER. When all three fields are selected, click the right arrow (>) to add them to the box on the right.

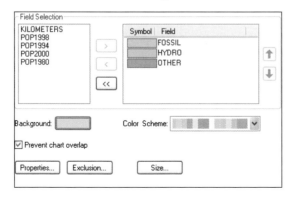

A symbol color is assigned to each attribute from the currently selected color scheme. You'll pick different colors.

17 In the Symbol column, right-click the FOSSIL symbol. In the color palette, click Mars Red. Right-click the HYDRO symbol and click Moorea Blue. Right-click the OTHER symbol and click Solar Yellow.

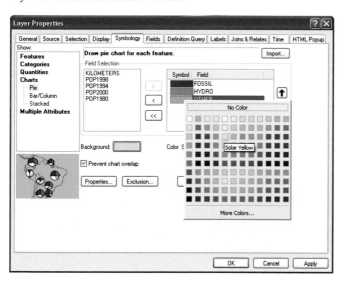

18 Click the Background color square to open the Symbol Selector.

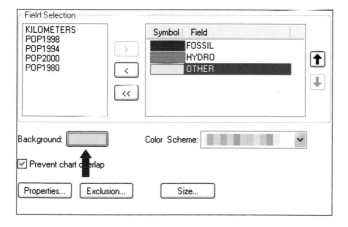

19 Click the Fill Color square. On the color palette, click Sahara Sand. Click the Outline Color square. On the color palette, click Gray 40%. Make sure your dialog box matches the following graphic, then click OK to return to the Layer Properties dialog box.

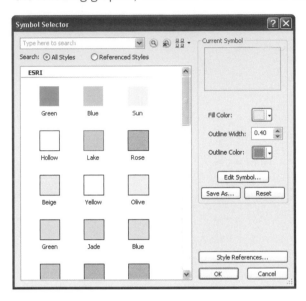

20 Near the bottom of the Symbology tab, click the Size button. The Pie Chart Size dialog box opens. Highlight the default size of 32 points and type **16**.

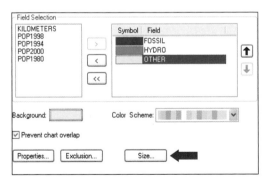 

21 In the Pie Chart Size dialog box, click OK. In the Layer Properties dialog box, click Apply. Move the Layer Properties dialog box away from the map.

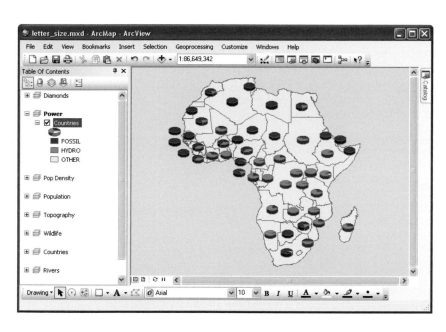

Each country is displayed with a 3D pie chart showing how much each energy source contributes to its production of electricity. On this map, the 3D effect is a bit overwhelming.

22 Near the bottom of the Layer Properties dialog box, click Properties to open the Chart Symbol Editor. Click the Display in 3D check box to uncheck it. Make sure your dialog box matches the following graphic, then click OK.

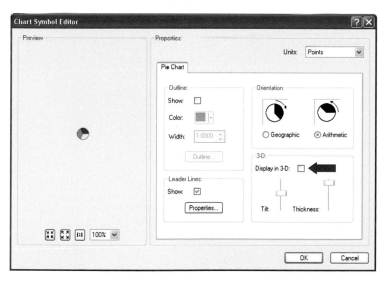

Removing the 3D effect from the chart symbols will make them seem a little too large. (Click Apply if you want to check this.)

23 On the Symbology tab, click Size. In the Pie Chart Size dialog box, highlight the current size of 16 points and type **12**. Click OK in the Pie Chart Size dialog box and click OK to close the Layer Properties dialog box.

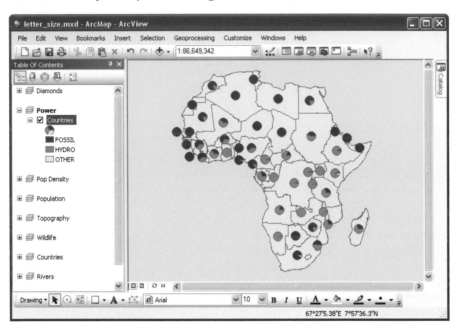

The map shows that in northern Africa (where oil is plentiful and water scarce) fossil fuels are the main source of electric power, while in central and southern Africa (where water is plentiful) power is generated chiefly by hydroelectricity.

It's time for a look at your finished poster.

24 Click the View menu and click Layout View.

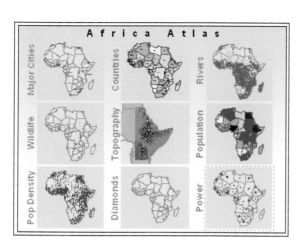

Your Africa Atlas poster is ready for Geography Awareness Week. In the meantime, you can send it to a plotter or printer and hang it on your wall.

In the current map document, the map page size is set to E, 44 inches wide by 34 inches high. (You can confirm this by clicking the File menu and clicking Page and Print Setup.) If you have access to a plotter, print this version. If not, you can make a letter-size version of the poster by printing the map document letter_size.mxd in the \GTKArcGIS\Chapter06\Data folder. Map elements in this document have been rescaled to fit a letter-size page. You'll find more information about printing maps in chapter 19.

25 If you want to save your work, click the File menu and click Save As. Navigate to **\GTKArcGIS\Chapter06\MyData**. Rename the file **my_ex06d.mxd** and click Save.

26 If you are continuing to the next chapter, leave ArcMap open. Otherwise, exit the application. Click No when prompted to save your changes.

Chapter 7

Labeling features

Using dynamic labels

Setting rules for placing labels

Using interactive labels and creating annotation

Broadly speaking, a label is any text that names or describes a feature on a map, including proper names (Thailand, Trafalgar Square, Highway 30), generic names (Tundra, Hospital), descriptions (Residential, Hazardous), or numbers (7,000 placed on a mountain to indicate its elevation). In ArcMap, labels specifically represent values in a layer attribute table. You can add other bits of text to a map, but, strictly speaking, they are not labels and do not behave quite the same. Labels, for example, stay close to the features they are associated with as you zoom in or out. Text is not similarly sensitive. Its position relative to features may change markedly if you zoom or pan the map.

As you have already seen in chapters 3 and 5, you can label all features in a layer with a single click. This process, called dynamic labeling, is fast but has some limitations. For one thing, ArcMap chooses the label positions. (You can set guidelines, but the actual placement is out of your hands.) If the scale of your map changes, or if different layers are added or removed, ArcMap may change the positions of labels.

Dynamically placed labels can, however, be converted to annotation. As annotation, each label becomes a piece of text that can be independently moved and modified. If you want significant control over the appearance and placement of labels, it is essential to convert them to annotation.

If you only need to label a few features on your map, the best solution may be the Label tool on the Draw toolbar. This tool allows you to label features one at a time by clicking them. Like annotation, interactively placed labels can be put wherever you want and changed individually.

Using dynamic labels

Dynamic labels are easy to work with because they behave as a group. When you turn them on or off, change their symbols, or change the attribute value they express, these operations are applied to all labels in the layer. The drawback to dynamic labels is that ArcMap doesn't always put them just where you'd like. For this reason, they are often converted to annotation in the late stages of map production.

In the following exercises, your map display will be redrawn many times as ArcMap adds and moves labels. If it fails to redraw as it should, click the Refresh View button at the bottom left corner of the map display.

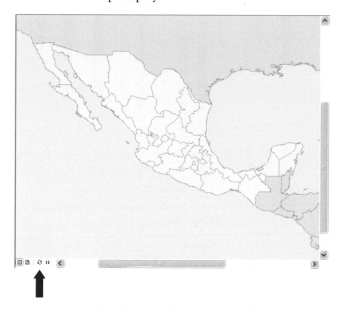

The positions ArcMap chooses for dynamic labels depend on many factors, including the size of your application window. Your results may be slightly different from those shown in the graphics.

Exercise 7a

You work for the Ministry of Tourism in Mexico and are making a map to promote travel in the southeastern states of Campeche, Chiapas, Quintana Roo, Tabasco, and Yucatán. The map will be published on the Internet, allowing map users worldwide to turn on various layers and navigate to different parts of the region, where beautiful beaches lie close to Mayan ruins. In this exercise, you'll label the five states and their major ruins. You'll also add text to the map.

1   Start ArcMap. In the ArcMap—Getting Started dialog box, under the Existing Maps section, click Browse for more. (If ArcMap is already running, click the File menu and click Open.) Navigate to **C:\ESRIPress\GTKArcGIS\Chapter07**. Click **ex07a.mxd** and click Open.

Mexico is shown in yellow. The five states that will be on your tourist map are on the Yucatán Peninsula in the southeast. You'll zoom in to them.

2   Click the Bookmarks menu and click Southeastern States.

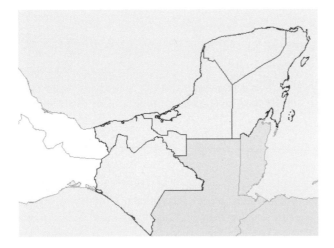

The Southeastern States layer displays only at scales larger than 1:20,000,000. You'll label the states in this layer dynamically.

3   In the table of contents, right-click the Southeastern States layer and click Label Features.

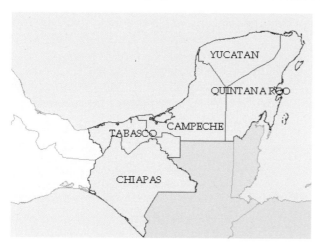

Each feature in the layer is labeled. ArcMap tries to put the labels in the center of each polygon. You'll change the font and color of the labels.

4   In the table of contents, double-click the Southeastern States layer to open the Layer Properties dialog box. Click the Labels tab.

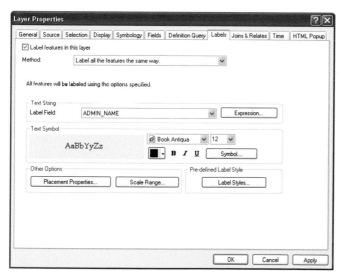

The Label Field drop-down list in the middle of the dialog box shows you which field in the attribute table is being used for labels (ADMIN_NAME). Below this, you see the current label symbology. The labels for this layer are drawn in black Book Antiqua 12 point. (If Book Antiqua is not one of your system fonts, another font, such as Arial, will be substituted.)

5   Click the Color square. On the color palette, click Gray 30%.

6   Click the drop-down arrow for setting the font. In the list of fonts, scroll up and click Arial.

7   In the Size drop-down list, click 12 and type **7** in its place.

8   Click the B (Bold) button. Make sure that your dialog box matches the following graphic, then click OK.

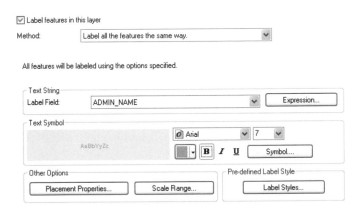

The new label properties are applied to the map.

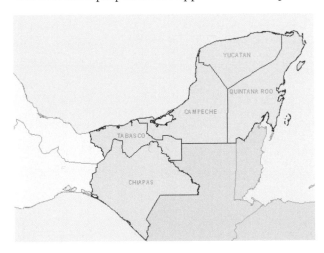

Map readers would be better oriented if you labeled the Gulf of Mexico in the north-west part of the map. Unfortunately, you don't have a layer that represents the ocean. (The blue water is simply the background color of the data frame.) Since there is no Gulf of Mexico feature, there is nothing to label. Instead, you'll add text with the New Text tool.

9 On the Draw toolbar, click the New Text tool.

10 Move the mouse pointer over the map. The cursor changes to a crosshair with an "A." Click in the Gulf of Mexico to add a text box.

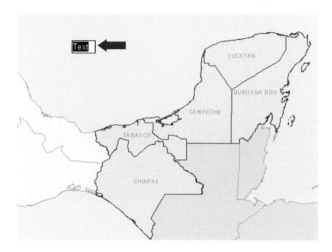

11 In the text box, type Gulf of Mexico and then press Enter.

The text is drawn in Arial 10 point, as you can see on the Draw toolbar. The dashed blue line indicates that the text is selected. You'll increase the font size.

12 Right-click "Gulf of Mexico." On the context menu, click Properties. On the Properties dialog box, click the Text tab, if necessary, and click Change Symbol.

The Symbol Selector dialog box opens.

13 In the Symbol Selector, scroll down until you see the predefined label style for Ocean, then click it. Make sure that your dialog box matches the following graphic, then click OK. Click OK again on the Properties dialog box to close it.

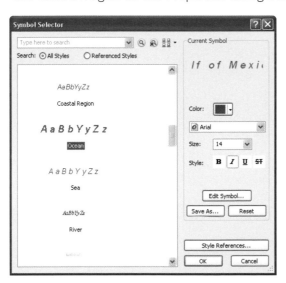

The new properties are applied to the text. Now that it's larger, the text may have to be moved.

14 Move the mouse pointer over the text. The cursor changes to a four-headed arrow. Drag the text so that it approximately matches the following graphic, then release the mouse button. Click an empty blue area to unselect the text.

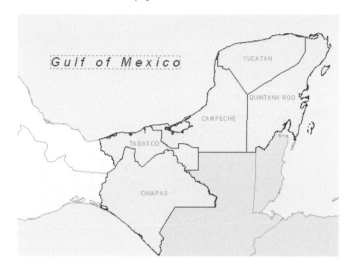

Now you will display the locations of major Mayan ruins.

15 In the table of contents, turn on the Ruins layer.

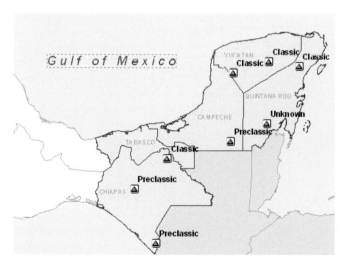

The labels show the period to which each ruin belongs—interesting, but less important for your map than the ruin's name. You will choose another field in the layer attribute table to use for labeling.

Notice that ArcMap has moved some of the state labels to new positions to make room for the ruin labels. (Again, your display may look different.)

16 In the table of contents, double-click the Ruins layer. In the Layer Properties dialog box, click the Labels tab. Click the Label Field drop-down arrow and click NAME. Make sure your dialog box matches the following graphic, then click OK.

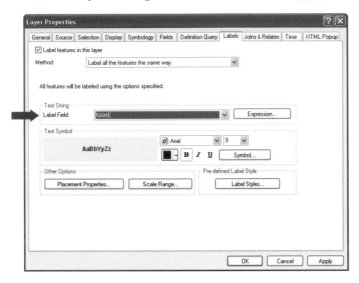

The ruins are now labeled with their names.

17 If you want to save your work, click the File menu and click Save As. Navigate to **\GTKArcGIS\Chapter07\MyData**. Rename the file **my_ex07a.mxd** and click Save.

18 If you are continuing with the next exercise, leave ArcMap open. Otherwise, exit the application. Click No if prompted to save your changes.

Setting rules for placing labels

When you place labels dynamically, ArcMap chooses label positions according to rules that you set. Labels for points can occupy any of eight positions around the feature (above, below, above right, below right, and so on). You can rank these positions and prohibit particular ones. Labels for line features can be placed above, below, or on the features, and can be made to follow their curves. Labels for polygons, however, don't have placement options—ArcMap places each label as close to the center of the polygon as it can.

In addition, labeling priorities can be set for each layer in a map. If there isn't room to label two nearby features in different layers, the feature in the layer with the higher priority will be labeled.

Exercise 7b

In this exercise, you will change the rule for positioning the ruin labels. You will also turn on and label the Rivers layer. Finally, you will change the layer priorities that determine which feature is labeled when there is a conflict.

1   In ArcMap, open **ex07b.mxd** from the **C:\ESRIPress\GTKArcGIS\Chapter07** folder.

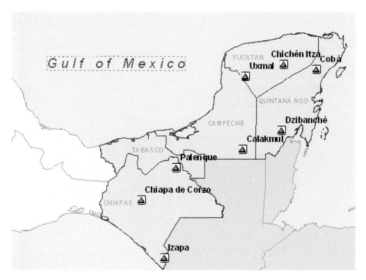

The map, zoomed in on the Yucatán region, looks as it did at the end of the last exercise. The ruin labels are not bad where they are, but might look better directly above the ruins.

2 In the table of contents, double-click the Ruins layer. In the Layer Properties dialog box, click the Labels tab.

3 Click Placement Properties. In the Placement Properties dialog box, click the Placement tab.

When labels are offset horizontally around points (the default), each possible position can be ranked. The graphic on the dialog box shows the current setting. The feature is represented by the white square in the middle and the numbers in the surrounding squares rank each position—"1" is the most desirable and "3" the least. A "0" means that a label may not occupy that position. In the current setting, ArcMap places labels above and to the right of features as long as it can do so without overlapping other labels in any layer. All positions are allowed, although positions below and to the left of the feature are discouraged.

You want to center the labels above the features.

4 Click Change Location to open the Initial point placement dialog box.

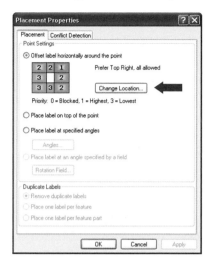

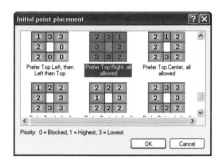

5 In the Initial point placement dialog box, scroll up and click the setting called Top Only, Prefer Center. Make sure that your dialog box matches the following graphic, then click OK. Click OK on the Placement Properties dialog box and click OK to close the Layer Properties dialog box.

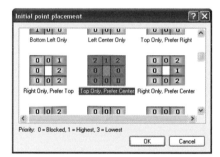

The ruin labels now appear directly above the features. If the map becomes crowded with other labels, the ruin labels will be shifted to the left or right of their current positions.

Now you'll turn on the Rivers and Lakes layers. When you do, ArcMap moves the Tabasco label to avoid overlapping a river.

6   In the table of contents, turn on the Rivers layer and the Lakes layer. If the Tabasco label does not redraw, click the Refresh View button in the lower left corner of the map display.

Before

After

7   In the table of contents, right-click the Rivers layer and click Label Features.

Only two rivers, Jalate and Azul, are labeled. (Again, depending on the size of your ArcMap window, your results may be different.) At this scale, more river labels can't be added without overlapping other labels.

Allowing labels to overlap

By default, ArcMap does not put labels on top of each other. This means that, depending on the scale and the part of the map you are looking at, some features in a layer may not be labeled. If you want to be sure that every feature in a layer is labeled, you can tell ArcMap to place overlapping labels. You can also control label-on-label and label-on-feature overlaps for individual layers. For more information, click the Contents tab in ArcGIS Desktop Help and navigate to *Professional Library > Mapping and Visualization > Adding text to a map > Displaying labels > Placing labels with the Standard Label Engine > Placing overlapping labels.*

To label more rivers, you'll use the Labeling toolbar to change the priority of the Rivers layer.

8   Click the Customize menu, point to Toolbars, and click Labeling.

The Labeling toolbar, with several buttons for managing labels, opens and floats above the ArcMap application window.

9   Click the Label Priority Ranking button.

The Label Priority Ranking dialog box opens.

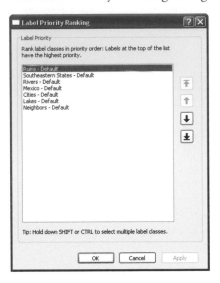

10 The layers in the map are listed in their order of labeling priority. You'll move the Rivers layer to the top.

In the Label Priority list, click Rivers–Default. Click the Top arrow to move the Rivers layer to the top of the list. Make sure your dialog box matches the following graphic, then click OK.

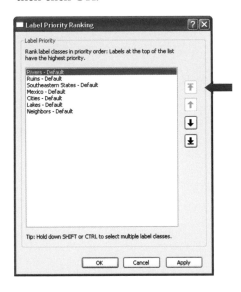

Exercise **7b** Setting rules for placing labels

Moving the Rivers layer to the top of the list does not guarantee that all rivers will be labeled, but at least one more river, Mezcalapa, should be labeled. (Notice that the Tabasco label has moved again—overlapping the Mezcalapa river feature is less problematic than overlapping the Mezcalapa label.)

You'll change the appearance of the river labels.

11 In the table of contents, double-click the Rivers layer. On the Layer Properties dialog box, click the Labels tab. Click Symbol, as shown in the following graphic.

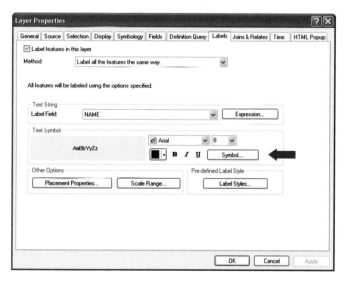

12 In the Symbol Selector, scroll down and click the River label style. Click the Bold button. Make sure that your dialog box matches the following graphic, then click OK. Click Apply on the Layer Properties dialog box.

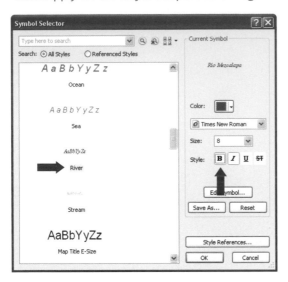

13 Move the Layer Properties dialog box away from the map. The new properties are applied to the river labels.

Another river, the Usumacinta, is labeled. Label display depends on the subtle interrelationship of features, other labels, and the display scale—even a small change in symbology can make labels appear or disappear.

Now you'll make the labels follow the curves of the rivers.

14 On the Labels tab, click Placement Properties.

15 In the Placement Properties dialog box, click the Placement tab. In the Orientation frame, click Curved.

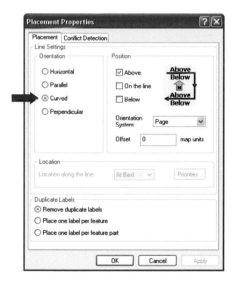

16 Click OK, then click OK again to close the Layer Properties dialog box.

The river labels now follow the curves of the features.

At the present display scale, the river labels are starting to clutter the map. You'll make them scale-dependent so that you don't see them until you zoom in.

17 In the table of contents, double-click the Rivers layer. Make sure the Labels tab is selected.

18 Click Scale Range to open the Scale Range dialog box.

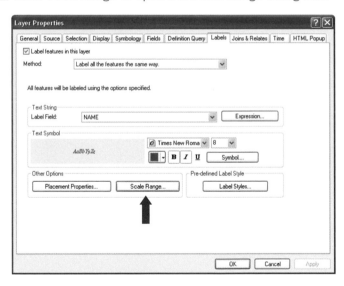

19 In the Scale Range dialog box, click the Don't show labels when zoomed option. In the Out beyond text box, type **3500000**. When focus is off the Out beyond text box, it turns into 1:3,500,000. Make sure your dialog box matches the following graphic, then click OK. Click OK in the Layer Properties dialog box to close it.

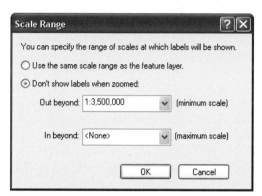

5

6

7

On the map, the river labels are no longer visible. They will be visible only at scales larger than 1:3,500,000. To see them, you'll zoom in to a bookmark.

20 Click the Bookmarks menu, and click Rivers.

Now when tourists zoom in, they'll find the rivers labeled. When they zoom out, the river labels won't clutter the map.

21 Close the Labeling toolbar.

22 If you want to save your work, save it as **my_ex07b.mxd** in the **\GTKArcGIS\Chapter07 \MyData** folder.

23 If you are continuing with the next exercise, leave ArcMap open. Otherwise, exit the application. Click No if prompted to save your changes.

Using interactive labels and creating annotation

Although you have considerable control over dynamically placed labels, you can't specify exact positions for them or symbolize them individually. To do that, you must convert the labels to annotation—this means, essentially, that you tell ArcMap to turn over full responsibility for label placement and appearance to you. Typically, you do this when your map is nearly finished and you want to make final adjustments to it.

Like other text and graphics, annotation is saved within your map document. If the data in your map is part of a geodatabase (such as the World.mdb geodatabase you worked with in chapter 4), you can also save annotation to the geodatabase. It then becomes much like any other spatial dataset: it can be copied, edited, and added to new map documents.

If you only need to label a few features in a layer, it is convenient to use the Label tool on the Draw toolbar. This tool lets you click the particular features you want to label and gives you the same full control over placement and appearance that you have with annotation.

For more information, click the Contents tab in ArcGIS Desktop Help and navigate to *Professional Library > Data Management > Geographic data types > Annotations.*

Exercise 7c

In this exercise, you'll label the city of Mérida. You'll also convert the labels for the Southeastern States layer to annotation so you can adjust their positions.

1   In ArcMap, open **ex07c.mxd** from the **C:\ESRIPress\GTKArcGIS\Chapter07** folder.

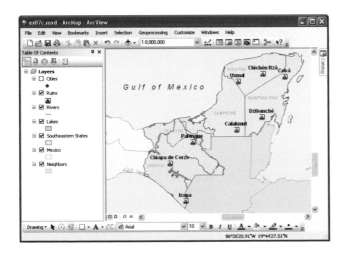

On the map, all layers except Cities are turned on.

2   In the table of contents, turn on the Cities layer. The map shows major cities in the region. (If the Yucatán label does not redraw, click the Refresh View button.)

The northernmost city on the peninsula is Mérida. As the largest city in the region, and the one most tourists will fly into, it should have a label.

3   On the Draw toolbar, click the drop-down arrow by the New Text tool. On the tool palette, click the Label tool.

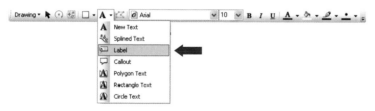

The Label Tool Options dialog box opens.

There are two placement options: you can let ArcMap choose the label position for you, or you can choose it yourself. You'll accept the default (Automatically find best placement) and adjust the position later.

4 Close the Label Tool Options dialog box. Locate the city of Mérida (map tips can help) and click to label it. If you miss and label the Yucatán region by mistake, press the Delete key, then click the Label tool and try again.

The label for Mérida may obscure the label for the ruin of Chichén Itzá. You'll move it to a new position.

5 On the Tools toolbar, click the Select Elements tool.

6 Move the mouse pointer over the Mérida label. The cursor changes to a four-headed arrow. Drag the label to a place just above and to the left of the point feature.

7 With the cursor over the Mérida label, right-click to open the context menu. On the context menu, point to Nudge and click Nudge Left. The label moves slightly to the left. You can also use the arrow keys on your keyboard to nudge the labels.

The labels for Mérida, Chichén Itzá, and Yucatán should all be visible. (If you don't see all three labels, click the Refresh button.)

8 Click somewhere in the blue area away from the Mérida label to unselect it.

Your map is almost finished. You'll convert the Southeastern States layer to annotation so you can make some final adjustments to the label positions.

9 In the table of contents, right-click the Southeastern States layer and click Convert Labels to Annotation. The Convert Labels to Annotation dialog box opens.

The default option for storing annotation is to store it in a geodatabase. Since you don't have your Mexico data in a geodatabase, however, you will simply store the annotation in the map document.

10 Click the option to store annotation in the map. Make sure your dialog box matches the following graphic, then click Convert.

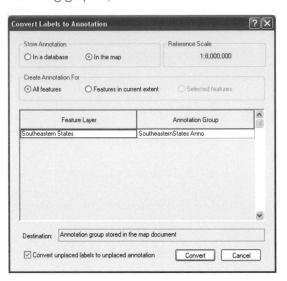

The labels are converted to annotation. In the process, their positions may change slightly and affect the display of other labels.

You'll move the labels for the states of Yucatán, Tabasco, and Chiapas to where they are less likely to be mistaken for the names of cities or ruins. Use the graphic following step 12 as a guide.

11 On the Tools toolbar, make sure the Select Elements tool is selected. Click the Yucatán label and drag it to the northeast corner of the state.

12 Click the Chiapas label and drag it slightly to the southeast. Click the Tabasco label and drag it west across the river. Click somewhere away from the label to unselect it. Click the Refresh View button if necessary.

Your map is ready for review at the Ministry of Tourism. One benefit of maps published on the Internet is that users can turn layers on and off according to their interests. You may eventually want to add and label such layers as roads, minor ruins, resorts, and campgrounds.

13 If you want to save your work, save it as **my_ex07c.mxd** in the **\GTKArcGIS\Chapter07 \MyData** folder.

14 If you are continuing with the next chapter, leave ArcMap open. Otherwise, exit the application. Click No if prompted to save your changes.

5
6
7

Chapter 8

Querying data

Identifying, selecting, finding, and hyperlinking features
Selecting features by attribute
Creating reports

A great strength of GIS is that it doesn't just show you where things are, it also tells you a lot about them. If you're curious about the road leading out of town, for example, you may be able to find out such things as its name, its length, its speed limit, and the number of lanes it has. If that road leads to a place like Carlisle, Pennsylvania, you might want to know the town's population (18,419), the median household income ($26,151), the median home value ($77,400), and many other things.

Much information about features can be conveyed by the way they are symbolized and labeled. But, for most features, it isn't possible to display everything that is known on a single map.

In ArcMap, there are several ways to retrieve unseen information about features. You can click features to display their attributes. This is called identifying features. You can click features to highlight them and look at their records in the layer attribute table. This is called selecting features interactively. You can write a query that automatically selects features meeting specific criteria (for example, three-bedroom houses with swimming pools). This is called selecting features by attributes. Or you can provide ArcMap with a piece of information, such as a name, and see which feature it belongs to. This is called finding features.

Attributes are not the only kind of information that can belong to a feature. Pictures, text documents, and Web pages can also be associated with features through hyperlinks. Clicking a hyperlinked feature opens a file on disk, points your Web browser to a URL, or runs a macro (a sequence of commands).

Identifying, selecting, finding, and hyperlinking features

The fastest way to get information about a single feature is to identify it. If you want to compare information about several features, the best way is to select the features on the map and look at their records in the layer attribute table. When you have a piece of information—a place name or address, for instance—but don't know which feature it belongs to, you can use the information to search the map for the feature.

Hyperlinks associate features with things that can't be stored as attribute values. A feature representing a government office, for example, could be linked to the office's Web page.

Exercise 8a

You are a real estate agent in Redlands, California. Many of your clients are new ESRI employees who have recently moved to the area. Your acquaintance with people in the industry helped you decide to use GIS in your own business.

Currently, you're working with a couple that has two children. They've asked to see properties in a neighborhood near the ESRI campus. They're looking for a three-bedroom house—preferably on a corner lot—and are willing to spend up to $175,000.

1 Start ArcMap. In the ArcMap—Getting Started dialog box, under the Existing Maps section, click Browse for more. (If ArcMap is already running, click the File menu and click Open.) Navigate to **C:\ESRIPress\GTKArcGIS\Chapter08**. Click **ex08a.mxd** and click Open.

The map shows local streets, land parcels in a neighborhood, and surrounding lots. The parcels are classified as either for sale or not for sale.

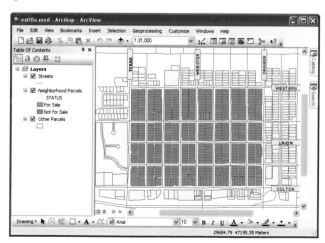

You will zoom in for a better look at the neighborhood and use the Identify tool to learn more about parcels that are for sale.

2   In the table of contents, right-click Neighborhood Parcels. On the context menu, click Zoom To Layer.

3   On the Tools toolbar, click the Identify tool.

4   Move the Identify window out of the way, if necessary. On the map, click one of the parcels for sale, such as the one in the following graphic.

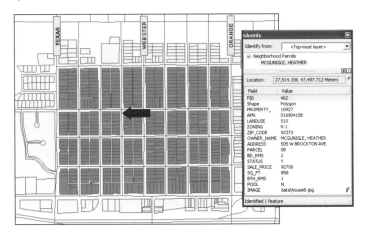

The bottom half of the Identify window displays the feature's attribute values. The box on the top tells you which layer is being identified (Neighborhood Parcels) and shows the value for the attribute that is set as the layer's primary display field (in this case, OWNER_NAME). If your window is not in a vertical orientation, click the Toggle Split Orientation button to change from a horizontal window orientation.

The Identify from drop-down list at the top lets you pick the layer whose features you are identifying. By default, this is the topmost layer. (The topmost layer is the first one in the table of contents that contains a feature at the location you click.)

5   Click a few more parcels for sale. The information in the Identify window changes.

Identifying features is the fastest way to get information about them, but it isn't convenient for comparing the attributes of several features. To do this, you'll select features on the map, then look at their records in the layer attribute table.

6   Close the Identify window. In the table of contents, click the List by Selection button.

By default, all layers are selectable. You want to select only features from the Neighborhood Parcels layer.

7   Beside the Streets and Other Parcels layers, click the toggle selectable button, as shown in the following graphic.

The Streets and Other Parcel layers are moved to a Not Selectable category.

8   Click the Selection menu and click Selection Options.

In the Selection Options dialog box, the selection tolerance is set to 5 pixels. Because you are working with small features that have many features adjacent to them, you'll reduce the tolerance. This ensures that when you click a location, you won't select more than one feature at a time.

9   In the Selection tolerance box, replace the value of 5 with **0** as shown in the following graphic, then click OK.

10   Click the Selection menu, point to Interactive Selection Method, and click Add to Current Selection.

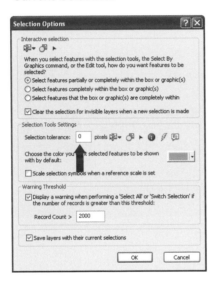

11   On the Tools toolbar, click the Select Features tool.

12   On the map, click each of the eight corner parcels that are for sale. Each parcel you

click is added to the selection set. If you select a parcel by mistake, click the Selection menu, point to Interactive Selection Method, and click Remove From Current Selection. Then click the selected parcel to unselect it. The table of contents shows how many parcels are selected and the names of the parcel owners.

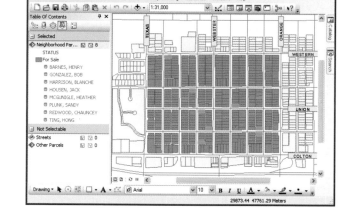

Selected parcels are outlined in cyan. In the attribute table for the layer, the records corresponding to the selected parcels are also highlighted.

13 Right-click Neighborhood Parcels and click Open Attribute Table. There are eight records highlighted. If not, click the Show selected records button.

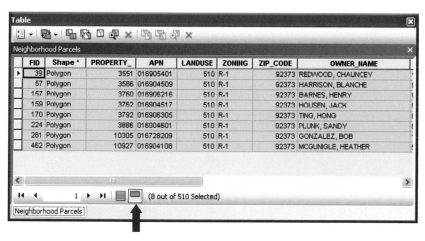

14 Scroll through the table to the right to see its attributes. You can now compare values like price or square footage for each house. Right-click the SALE_PRICE field name. On the context menu, click Sort Descending.

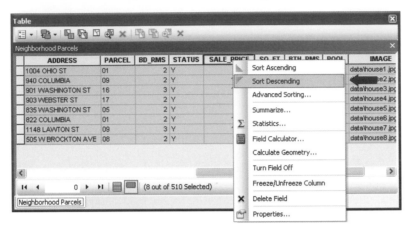

Now the records are sorted in order from most expensive to least expensive. The first record shows a value of 166500 (or $166,500).

15 Click the gray tab to the left of the first record (the house at 901 Washington St.).

The record is highlighted in yellow as is the corresponding map feature. You may need to move the attribute table to see the highlighted feature.

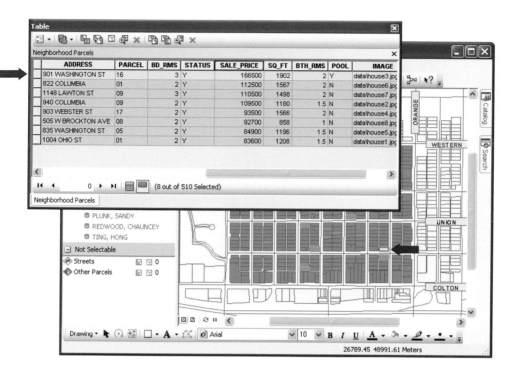

16 Close the attribute table. The yellow highlight on the feature disappears. Click the Selection menu and click Clear Selected Features.

Changing the selection color for features and records

The default selection colors can be changed. They can also be set independently for each layer in a map. For more information, click the Contents tab in ArcGIS Desktop Help and navigate to *Professional Library > Data Management > Geographic data types > Tables > Displaying tables > Using field properties, aliases, and table display options > Understanding field properties, aliases, and table display options*.

Attribute information is useful, but a photo conveys information that table attributes can't—such as whether a house is attractive to you. Pictures, documents, Web pages, and macros can be hyperlinked to features.

The attribute table of the Neighborhood Parcels layer has a field called IMAGE that contains paths to photographs of the three-bedroom corner houses that are for sale. You will hyperlink the features to the photos.

17 In the table of contents, double-click Neighborhood Parcels. In the Layer Properties dialog box, click the Display tab.

18 Check Support Hyperlinks using field. Click the drop-down arrow and click IMAGE. Below the drop-down list, make sure the Document option is selected, as shown in the following graphic. Click OK.

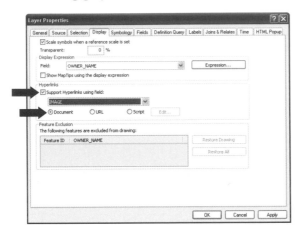

Now your clients can see photos of the properties before deciding whether to visit them.

19 On the Tools toolbar, click the Hyperlink tool.

On the map, hyperlinked features are outlined in blue.

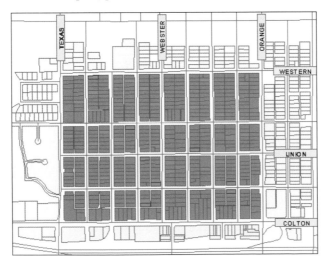

20 Click any hyperlinked feature to display a photo of the property. The photo opens in your default image-browsing software. After you look at it, close the application that opened it.

Creating hyperlinks

There are two ways to create hyperlinks. One is to add document paths or URL addresses to a field in the layer attribute table. This is efficient if you are hyperlinking many features. The other way is to click a feature and specify the document path or URL. This is easier if you are setting just a few hyperlinks. It is also the only way that lets you hyperlink a feature to more than one object. For more information, click the Contents tab in ArcGIS Desktop Help and navigate to *Professional Library > Mapping and Visualization > Working with layers > Interacting with layer contents > Using hyperlinks*.

When you take your clients on a tour of the neighborhood, they see a house they like in spite of its not being on a corner. You jot down the address (831 Washington St.) so you can get more information back at the office. You'll use the Find tool to locate the house on the map and display its attributes.

21 On the Tools toolbar, click the Find tool to open the Find dialog box.

22 In the Find dialog box, click the Features tab. Type **831 Washington St** in the Find box. (Do not type a period after the "St" abbreviation.)

To speed things up, you can specify a layer, and even a field, to search in.

23 Click the In drop-down arrow and click Neighborhood Parcels. In the Search options, click the In field option, then click the drop-down arrow and click ADDRESS. Make sure your window matches the following graphic, then click Find.

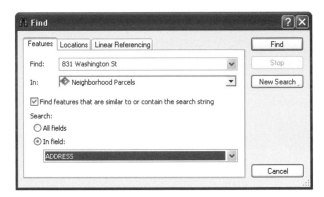

ArcMap searches the ADDRESS field of the Neighborhood Parcels attribute table for "831 Washington St" and displays the matching record.

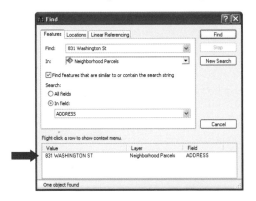

Now that you've found the feature, you can locate it on the map and get its attributes.

24 Move the Find window away from the map but where you can still see it. Click 831 WASHINGTON ST at the bottom of window.

The parcel flashes briefly in the lower right corner of the map. (If the Identify window covers the parcel, move it and repeat step 24.)

25 On the Tools toolbar, click the Identify button and click the 831 Washington St parcel.

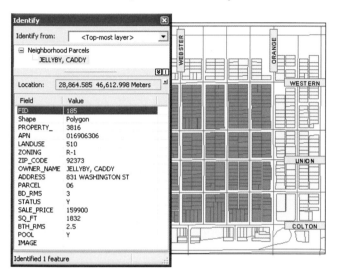

26 In the Identify window, scroll down to find the price and number of bedrooms. Scan the other attribute values. When you finish, close the Identify window and the Find window.

The house has three bedrooms, two-and-a-half bathrooms, and a swimming pool for an asking price of $159,900. It sounds great, and your clients are now willing to consider other houses that aren't on corners.

27 If you want to save your work, click the File menu and click Save As. Navigate to **\GTKArcGIS\Chapter08\MyData**. Rename the file **my_ex08a.mxd** and click Save.

28 If you are continuing with the next exercise, leave ArcMap open. Otherwise, exit the application. Click No if prompted to save your changes.

Selecting features by attribute

Interactive selection works when you can see what you're looking for on the map—corner lots, for instance. To select features according to criteria that you can't see on the map (such as having three bedrooms and a sale price under $175,000), you write a query.

A query selects features that meet specified conditions. The simplest query consists of an attribute (such as number of bedrooms), a value (such as three), and a relationship between them (such as "greater than" or "equal to"). Complex queries can be created by connecting simple queries with terms like "and" and "or." Relationship terms and connecting terms are called operators.

Queries are not written in ordinary English but in Structured Query Language (SQL). This is not as daunting as it might sound. To build a query, all you need to do is open a dialog box and click an attribute, an operator, and a value. There is also a Query Wizard that lets you write queries by choosing from drop-down lists.

Exercise 8b

Although your clients like the house at 831 Washington, they see no reason not to search a little further. You will help them select all three-bedroom houses in the neighborhood that have a sale price under $175,000.

1   In ArcMap, open **ex08b.mxd** from the **C:\ESRIPress\GTKArcGIS\Chapter08** folder.

The map of neighborhood parcels displays.

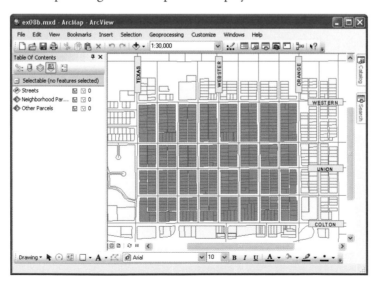

2   Click the Selection menu and click Select by Attributes. In the Select By Attributes dialog box, click the Layer drop-down arrow and click Neighborhood Parcels.

The fields from the parcels attribute table appear in the Fields list near the top of the dialog box. When a field is highlighted, you can display its attribute values in the Unique Values list on the right. The buttons on the left are used to choose operators.

The first condition you want to test is whether a house is for sale. Houses for sale have a STATUS of "Y."

3   In the Fields box, scroll to "STATUS" and double-click it. Click the equals (=) button. Click Get Unique Values. In the Unique Values box, double-click "Y." Your query is displayed in the expression box at the bottom of the dialog box.

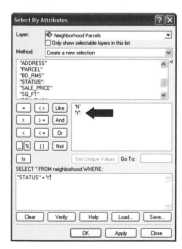

When you apply the query, ArcMap searches the attribute table for records that have "Y" in the STATUS field. The corresponding features are selected on the map.

4   Click Apply and move the dialog box out of the way. On the map, parcels that are for sale are selected.

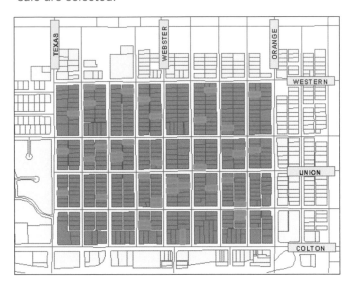

You already knew which parcels were for sale from the layer symbology. You'll make the query more complex so that only three-bedroom houses for sale are selected.

5   In the Select By Attributes dialog box, click the And button. In the Fields box, double-click "BD_RMS". Click the equals (=) button. Click Get Unique Values, then double-click 3 in the Unique Values box. Your query should match the following graphic.

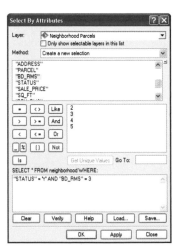

Records will be selected only if they have "Y" in the STATUS field and 3 in the BD_RMS field.

6 Click Apply. On the map, fewer houses are selected because fewer satisfy both conditions.

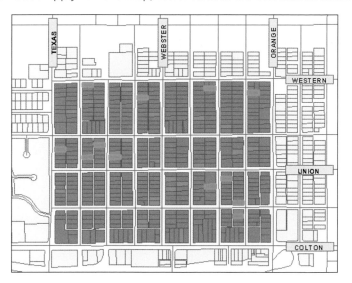

You will add one more condition to the query.

7 In the Select By Attributes dialog box, click the And button. In the Fields box, double-click "SALE_PRICE". Click the Less than (<) button. In the expression box, press the space bar and type **175000**. Your query should match the following graphic.

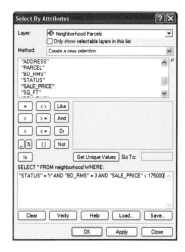

In the attribute table, only records with "Y" in the STATUS field, 3 in the BD_RMS field, and a number under 175000 in the SALE_PRICE field will be selected.

8 Click Apply. On the map, one of the houses is unselected because its price was too high.

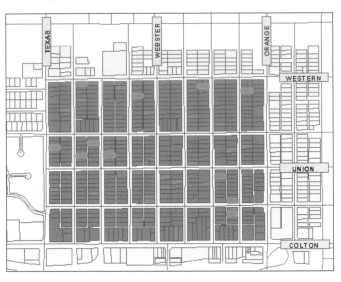

Eleven parcels are selected, including the house at 831 Washington and two corner lots from the original list. In the next exercise, you will make a report of the query results that your clients can take with them.

9 In the Select By Attributes dialog box, click Clear to remove the query from the box. Click Close.

10 If you want to save your work, save it as **my_ex08b.mxd** in the **\GTKArcGIS\Chapter08 \MyData** folder.

11 If you are continuing with the next exercise, leave ArcMap open. Otherwise, exit the application. Click No if prompted to save your changes.

Creating reports

Reports let you organize, format, and print the information contained in an attribute table. ArcMap's built-in report generator, which you'll use in this exercise, allows you to create simple reports that can be added to a map layout. For more detailed reports, you can use Business Objects Crystal Reports. This report-creation software comes with ArcGIS Desktop and is optional to install.

Exercise 8c

In the previous exercise, you selected eleven houses that met your clients' conditions. In this exercise, you will create a report they can take when they go to look at these houses.

1 In ArcMap, open **ex08c.mxd** from the **C:\ESRIPress\GTKArcGIS\Chapter08** folder.

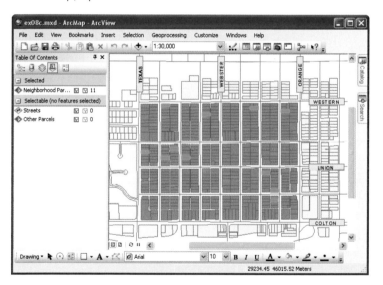

The map with eleven selected parcels displays.

2 Click the View menu, point to Reports, and click Create Report. The Report Wizard dialog box opens.

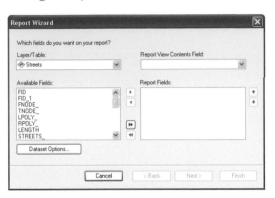

The Layer/Table drop-down list specifies the layer or table you will work with. The Available Fields scrolling box lets you choose the attributes to include in the report.

Your clients want to have street addresses and information on sale price, square footage, and number of bathrooms. You don't need to include the number of bedrooms—all the houses are known to have three.

3 Make sure that the Layer/Table drop-down list shows Neighborhood Parcels. In the Available Fields box, click ADDRESS. Click the Add fields button (right arrow) in the middle of the dialog box. The ADDRESS field moves from the Available Fields box to the Report Fields box.

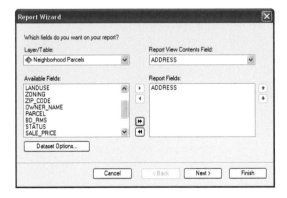

The ADDRESS field now appears in the Report Viewer Contents Field drop-down list. That means it will be displayed when you preview the report.

4 In the Available Fields box, double-click SALE_PRICE to add it to the Report Fields box. Do the same for the SQ_FT and BTH_RMS fields.

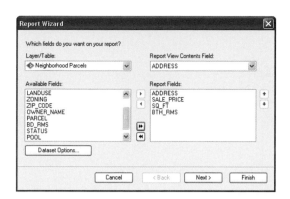

5 Since you want the report to contain data only for the eleven selected parcels, click the Dataset Options button and choose the Selected Set option.

6 Click OK and click Next in the Report Wizard dialog box. Click Next again. You will not be adding any grouping levels.

7 In the Fields column, choose SALE_PRICE from the drop-down list and in the Sort column, choose Ascending from the drop-down list.

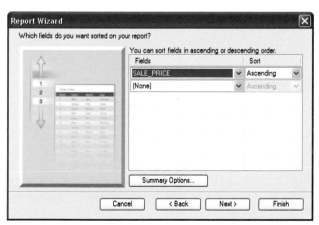

In the report, the houses will appear from least expensive to most expensive.

The size of your dialog box may be different, but you probably can't see the whole report. The drop-down list at the top of the dialog box lets you change the preview scale. (It doesn't change the printed size of the report.)

8 Click Next. Set the orientation to Landscape, then click Next.

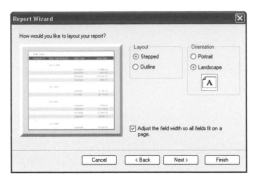

9 Make sure the Havelock (Default) style is selected, then click Next.

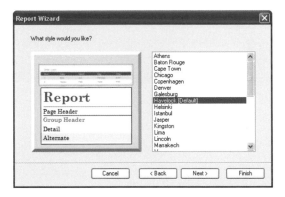

10 Replace the text in the title box with **Available Three Bedroom Houses**. Make sure the Preview Report option is chosen and click Finish.

The report is previewed. If you like the way the report looks you could click the Print button to send it to your printer.

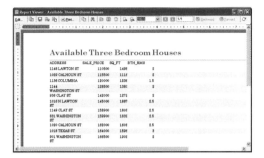

Most reports, including this one, need additional work. For example, the street names, numbers, and types in the address column each take up two lines. Your clients might appreciate having the listings better organized and aligned. In the next steps you will give the report a professional look by setting a few of its properties.

11   Click the Edit button in the top left corner of the Report Viewer window.

The window changes to the Report Designer window. The top right box lists the report components you can customize. Below it are the element properties associated with the highlighted component and the current value of each property. When a report component is highlighted, the element is selected in the report template view in the middle. The template area represents how your report pages will look.

To fit each address on one line, you need to widen the report's Address column.

12   In the report components list, expand detail (click the plus sign beside it) and click txtADDRESS to highlight it. In the Element Properties, scroll down to the Layout category and expand the Size property. The current size is shown as 1.208, 0.311 in.

13 In the Width property, double-click its current value and replace the current value with **3** to allow three inches for addresses on the printed page. Press Enter.

14 Click the Run Report button to preview your change.

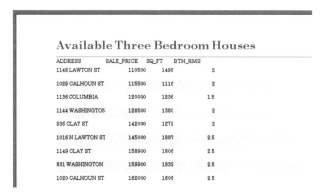

The addresses are now on one line and are easier to read. But the headings are still not aligned properly. Click the Edit button again to return the design mode.

15 In the template area, click the SALE_PRICE header component. In the Element Properties, under the Appearance category and change the Alignment property to Right.

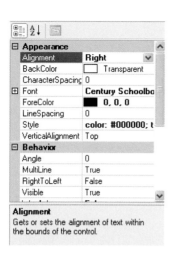

16 Repeat the same right alignment for the SQ_FT and BTH_RMS header components and click the Run Report button to preview the report one more time.

Available Three Bedroom Houses

ADDRESS	SALE_PRICE	SQ_FT	BTH_RMS
1148 LAWTON ST	110500	1498	2
1029 CALHOUN ST	115500	1116	2
1136 COLUMBIA	120000	1236	1.5
1144 WASHINGTON	128500	1880	2
936 CLAY ST	142000	1272	2
1016 N LAWTON ST	145000	1887	2.5
1149 CLAY ST	158900	1808	2.5
831 WASHINGTON	159900	1832	2.5
1020 CALHOUN ST	162000	1608	2.5

The SALE_PRICE, SQ_FT and BTH_RMS headings are now aligned properly with their column values.

17 At the top of the Report Viewer window, click the Print button.

18 In the Print dialog box, click Print to print the report.

19 If you want to save the report, click the Save Report button in the Report Viewer. In the Save Report Document dialog box, navigate to **\GTKArcGIS\Chapter08\MyData**. Name the file **my_report** and click Save. If you don't want to save the report, click No.

Adding reports to layouts

Within a map document, you can add a report to a layout by clicking the Add button in the Report Viewer dialog box. If you have saved a report to disk, you can add it to any map layout by loading it from the Report Properties dialog box. For more information, click the Contents tab in ArcGIS Desktop Help and navigate to *Professional Library > Mapping and Visualization > Reports > Saving, exporting, loading and running and a report > Loading a report.*

There is no need to save the map document. You haven't made any changes to it.

20 If you are continuing with the next chapter, leave ArcMap open. Otherwise, exit the application. Click No if prompted to save your changes.

Chapter 9

Joining and relating tables

Joining tables

Relating tables

Layer attribute tables contain descriptions of features (like the name of a town, the price of a house, or the length of an airplane flight), and they also contain spatial information, which is what enables ArcMap to draw a town, a house, or a flight path in the right place on a map. Spatial information, stored in the Shape field, specifies the type of feature in a layer (point, line, or polygon) and defines the location of each individual feature.

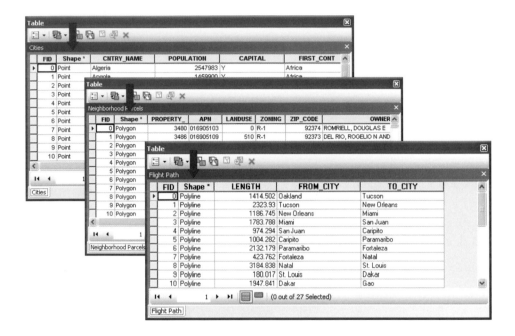

The Shape field of a layer attribute table specifies the feature type. It also stores each feature's geographic location. You don't see the location information when you look at the table, but it's there all the same.

The descriptive information in a layer attribute table, however, doesn't necessarily include everything you might want to show on a map. For example, the table below contains just a few basic facts about world countries.

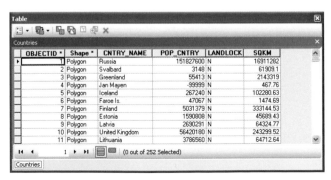

Suppose that you wanted to make a world map of per capita GDP or average life expectancy? You would need some attributes like those found in the next table of socioeconomic data.

OID	COUNTRY	BIRTH RATE	DEATH RATE	LIFE EXPECT (F)	LIFE EXPECT (M)	LITERACY RATE	PER CAPITA GDP
2	Afghanistan	41.82	18.01	45.1	46.62	31.5%	$792.00
3	Algeria	23.14	5.3	71.02	68.34	61.6%	$4,531.00
4	Azerbaijan	18.08	9.47	67.45	58.51	97%	$1,704.00
5	Albania	19.47	6.5	74.59	68.75	93%	$1,530.00
6	Armenia	10.97	9.53	71.04	61.98	99%	$2,777.00
8	Angola	46.89	25.01	39.56	37.11	42%	$1,099.00
10	Argentina	18.59	7.59	78.61	71.67	96.2%	$9,453.00
11	Australia	13.08	7.12	82.74	76.9	100%	$20,643.00

Unlike the previous table, this one has lots of attributes, but no Shape field. In other words, it stores descriptions but not spatial information. It's the sort of table you might find in a non-GIS database, or that you might even create yourself from facts based on your own research. By itself, this table can't be used to map data because its records have no relationship to features.

The question then becomes, how do you connect a nonspatial table like this to a layer attribute table? You can do it in either of two ways: with a table join or a table relate. A join appends the attributes of the nonspatial table to the layer attribute table, making one big table. A relate keeps the two tables separate but linked, so that record selections in one table cause corresponding selections in the other.

In either case, the process depends on making record matches between the nonspatial table and the layer attribute table. For records to be matched, the tables need a common attribute; that is, a field of shared values. In the example above, the Country field would be a common attribute, since both tables have a field of country names. Once the common attribute has been specified, then the Afghanistan record in the nonspatial table (along with all its attributes) can be matched to the Afghanistan record in the layer attribute table. The same goes for Algeria, Azerbaijan, and the rest.

Whether you join or relate tables depends on a relationship called cardinality between the tables. You join tables when each record in the layer attribute table has no more than one matching record in the nonspatial table.

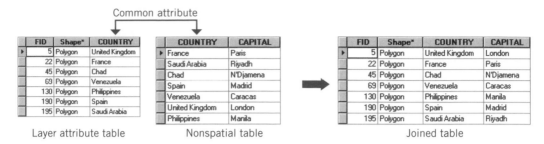

Common attribute

Layer attribute table				Nonspatial table				Joined table			

Layer attribute table:

FID	Shape*	COUNTRY
5	Polygon	United Kingdom
22	Polygon	France
45	Polygon	Chad
69	Polygon	Venezuela
130	Polygon	Philippines
190	Polygon	Spain
195	Polygon	Saudi Arabia

Nonspatial table:

COUNTRY	CAPITAL
France	Paris
Saudi Arabia	Riyadh
Chad	N'Djamena
Spain	Madrid
Venezuela	Caracas
United Kingdom	London
Philippines	Manila

Joined table:

FID	Shape*	COUNTRY	CAPITAL
5	Polygon	United Kingdom	London
22	Polygon	France	Paris
45	Polygon	Chad	N'Djamena
69	Polygon	Venezuela	Caracas
130	Polygon	Philippines	Manila
190	Polygon	Spain	Madrid
195	Polygon	Saudi Arabia	Riyadh

Each record in the layer attribute table has just one match in the nonspatial table (each country has just one capital). It so happens that the reverse is also true—each capital belongs to just one country. Another way to express this relationship is to say that the table cardinality is one-to-one.

You relate tables when each record in the layer attribute table may have more than one match in the nonspatial table, as in the next example. When you select France in the layer attribute table, all the matching records in the nonspatial table are also selected.

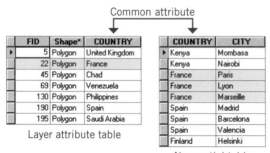

Common attribute

Layer attribute table:

FID	Shape*	COUNTRY
5	Polygon	United Kingdom
22	Polygon	France
45	Polygon	Chad
69	Polygon	Venezuela
130	Polygon	Philippines
190	Polygon	Spain
195	Polygon	Saudi Arabia

Nonspatial table:

COUNTRY	CITY
Kenya	Mombasa
Kenya	Nairobi
France	Paris
France	Lyon
France	Marseille
Spain	Madrid
Spain	Barcelona
Spain	Valencia
Finland	Helsinki

Here, the table cardinality is one-to-many. Each record in the layer attribute table has many matches in the nonspatial table (because there are many cities in a country). By joining the tables, you would lose information. For example, only one of the three matching records for France in the nonspatial table could be appended to the layer attribute table.

Joining tables

Once tables are joined, you can use the appended attributes of the nonspatial table to symbolize, label, query, and analyze the features in a layer.

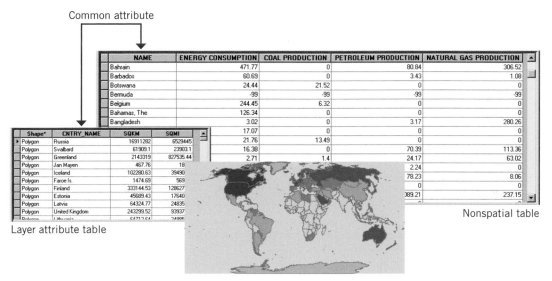

Common attribute

Layer attribute table

Nonspatial table

When a nonspatial table of world energy statistics is joined to a layer attribute table of countries, features can be symbolized by energy attributes. Here, a graduated color scheme of light orange to red shows per capita energy consumption in millions of BTUs.

A table join is preserved only within a map document—the tables remain separate on disk—and can be undone at any time. You can permanently append attributes by exporting a joined layer as a new dataset. You'll learn how to export a layer in chapter 11.

Exercise 9a

Between 1993 and 1998, abandoned oil extraction sites (called pits) in coastal Louisiana were evaluated for their risk to the environment. The pits, dating back to 1906, include wells, storage tanks, and other equipment and structures that may leak oil and possibly contaminate nearby waterways. If these structures are exposed by erosion, as often happens, they may be struck by boats, risking an oil spill. During the study, each pit was given a hazard rating and most were assigned specific recommendations for cleanup.

As a GIS analyst for the state of Louisiana, you'll make a map that symbolizes the most hazardous pits according to the type of cleanup that has been recommended for them. The map will show how many sites need each kind of cleanup and which parishes (counties) have the most serious problems. The map will help lawmakers decide which sites to clean up first, and how much money needs to be spent in each parish to do the job.

1 Start ArcMap. In the ArcMap—Getting Started dialog box, under the Existing Maps section, click Browse for more. (If ArcMap is already running, click the File menu and click Open.) Navigate to **C:\ESRIPress\GTKArcGIS\Chapter09**. Click **ex09a.mxd** and click Open.

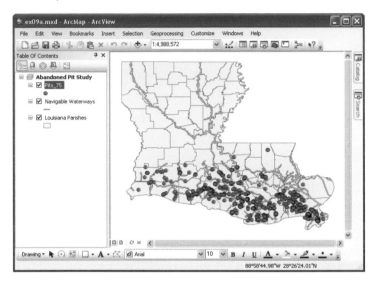

You see a map of Louisiana showing parishes, navigable waterways, and pits.

Southern Louisiana has more than 24,000 abandoned pits. Each has been given a hazard rating from 0 to 90. The Pits_75 layer in the table of contents has 606 pits, each with a hazard rating of 75 or higher.

2 Click the Bookmarks menu and click Southern Louisiana. The view zooms to the southern part of the state.

3 Right-click the Pits_75 layer and click Open Attribute Table. If necessary, click the Show all records button.

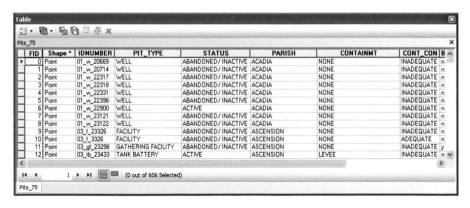

The IDNUMBER attribute identifies each pit, and PARISH names the parish it's found in. Other attributes provide detailed information about the pits, but there is no attribute for the recommended cleanup. This information is stored in a nonspatial table that you'll join to the Pits_75 layer.

4 Close the Pits_75 attribute table. On the Standard toolbar, click the Add Data button.

5 In the Add Data dialog box, navigate to **\GTKArcGIS\Chapter09\Data**. Click **Remedial_actions.dbf**, as shown in the following graphic, then click Add.

The Remedial_actions table is added to the table of contents. When a table is added to a map, the button at the top of the table of contents switches from List by Drawing Order to List By Source. With the List By Source button enabled, paths to disk are shown for all data in the map. In this case, since all the data comes from the same folder, only one path is displayed. (The table of contents is shown here as floating for example purposes, yours may be docked.)

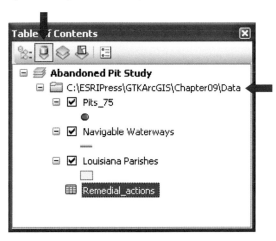

6 In the table of contents, right-click the Remedial_actions table and click Open.

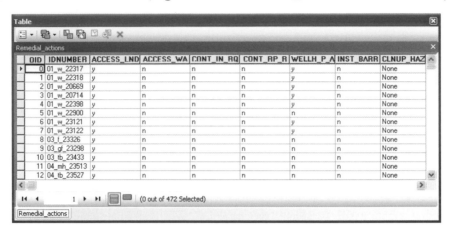

The IDNUMBER attribute identifies pits and is the common attribute that will be used to join the tables. (The field name of the common attribute is the same in both tables, but it doesn't have to be. What matters is that the values identify the same pits; for example, that Pit 01_w_22317 in the Remedial_actions table is the same pit as Pit 01_w_22317 in the Pits_75 table.)

7 In the Attributes of Remedial_actions table, right-click the IDNUMBER field heading. On the context menu, click Freeze/Unfreeze Column.

8 Scroll all the way to the right. The IDNUMBER field does not move.

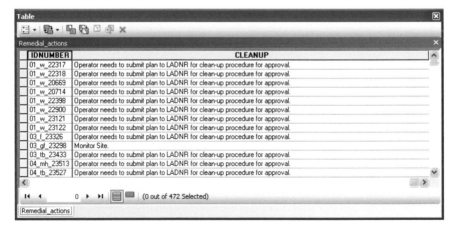

The recommended cleanup is the last field in the table.

9 Scroll down through the table.

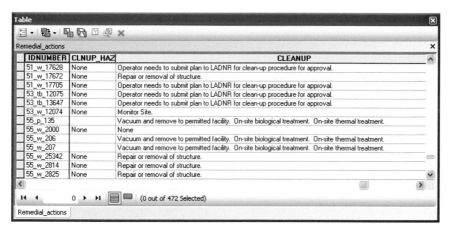

For many sites, a cleanup plan has yet to be submitted. Other sites have specific recommendations: continued monitoring, repair or removal, and on-site treatment.

Both the Remedial_actions table and the Pits_75 layer attribute table contain one record for each pit. The record relationship between the tables is therefore one-to-one. For each record in the Pits_75 table, there is no more than one matching record in the Remedial_actions table.

The Remedial_actions table has fewer records (472) than the Pits_75 table (606). It may be that cleanup recommendations have not yet been made for the other 134 pits. Records that don't have matches in the Remedial_actions table will have <Null> attribute values appended to them.

10 Close the Attributes of Remedial_actions table. In the table of contents, right-click the Pits_75 layer. On the context menu, point to Joins and Relates, then click Join.

11 In the drop-down list at the top of the Join Data dialog box, make sure "Join attributes from a table" is selected. In drop-down list number 1, click the drop-down arrow and click IDNUMBER. This specifies IDNUMBER as the common attribute in the Pits_75 table.

12 In drop-down list number 2, make sure the Remedial_actions table is chosen. In number 3, make sure the IDNUMBER field is chosen. This specifies IDNUMBER as the common attribute in the Remedial_actions table. In the Join Options section, you want to keep all the records. Make sure your dialog box matches the following graphic, then click OK.

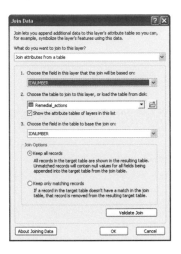

A message appears. ArcMap wants to know if it should build an index for the IDNUMBER field in the Remedial_actions table.

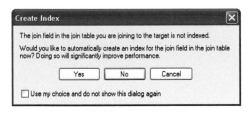

13 Click Yes to build the index.

ArcMap joins the attributes in the Remedial_actions table to the Pits_75 table, matching the records by their ID numbers.

14 In the table of contents, right-click the Pits_75 layer and click Open Attribute Table.

The joined table contains the attributes of both the Pits_75 table and the Remedial_actions table. Scroll to the right until you see IDNUMBER *. (The * indicates that the field has been indexed.) To the right of IDNUMBER * are the fields from the Remedial_actions table and are shown using field aliases. Field aliases are a more "user friendly" description of the content of the field.

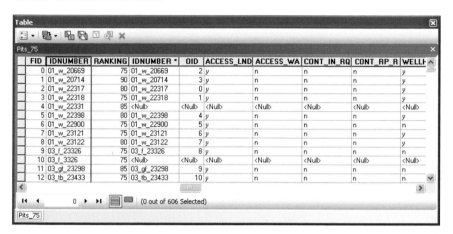

15 To turn off the field aliases, click the Table Options menu and click Show Field Aliases on the context menu.

The field names with the Remedial_actions prefix identifies that the attribute comes from the Remedial_actions table. Resize the fields as necessary to see its entire name.

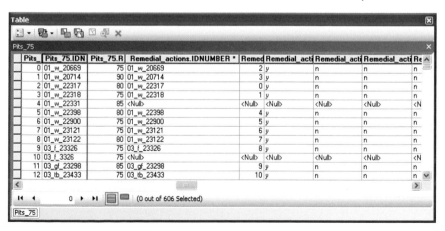

16 Scroll all the way to the right. The Remedial_actions.CLEANUP field contains the recommended cleanup for each pit.

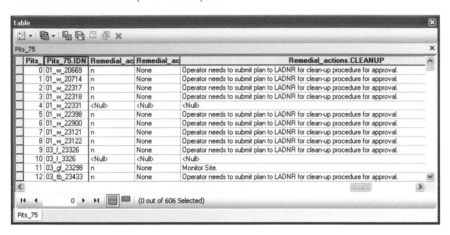

Pits_	Pits_75.IDN	Remedial_ac	Remedial_ac	Remedial_actions.CLEANUP
0	01_w_20669	n	None	Operator needs to submit plan to LADNR for clean-up procedure for approval.
1	01_w_20714	n	None	Operator needs to submit plan to LADNR for clean-up procedure for approval.
2	01_w_22317	n	None	Operator needs to submit plan to LADNR for clean-up procedure for approval.
3	01_w_22318	n	None	Operator needs to submit plan to LADNR for clean-up procedure for approval.
4	01_w_22331	<Null>	<Null>	<Null>
5	01_w_22398	n	None	Operator needs to submit plan to LADNR for clean-up procedure for approval.
6	01_w_22900	n	None	Operator needs to submit plan to LADNR for clean-up procedure for approval.
7	01_w_23121	n	None	Operator needs to submit plan to LADNR for clean-up procedure for approval.
8	01_w_23122	n	None	Operator needs to submit plan to LADNR for clean-up procedure for approval.
9	03_f_23326	n	None	Operator needs to submit plan to LADNR for clean-up procedure for approval.
10	03_f_3326	<Null>	<Null>	<Null>
11	03_gf_23298	n	None	Monitor Site.
12	03_tb_23433	n	None	Operator needs to submit plan to LADNR for clean-up procedure for approval.

(0 out of 606 Selected)

Pits_75

The values are <Null> wherever a pit does not have a matching record in the Remedial_actions table.

17 Close the table. In the table of contents, double-click the Pits_75 layer to open the Layer Properties dialog box. Click the Symbology tab.

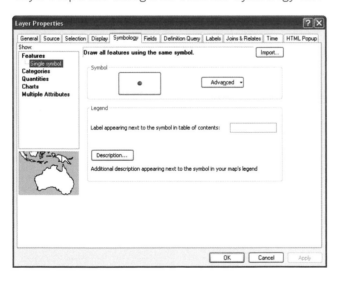

18 In the Show box, click Categories. Click the Value Field drop-down arrow, scroll to the bottom of the list, and click CLEANUP. Click the Add All Values button.

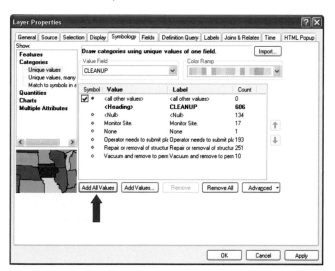

In the Value column, the five cleanup procedures are listed, though some of the descriptions are cut off. (You can see the full description as a map tip by placing your mouse pointer over the value.) The Count column shows the number of features corresponding to each type of cleanup.

19 Click the Symbol column heading and click Properties for All Symbols. The Symbol Selector dialog box opens. In the scrolling box, click Circle 2. Highlight the default size and type **8**. Make sure your dialog box matches the following graphic, then click OK.

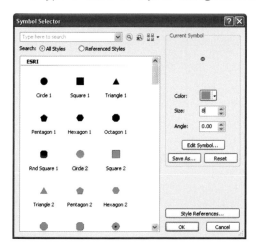

20 On the Symbology tab, uncheck the <all other values> check box. Click the Color Ramp drop-down arrow and click the Basic Random scheme, the first one in the list. Make sure your dialog box matches the following graphic (your colors may be different) and click OK.

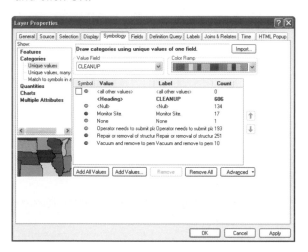

The symbology is applied to the map.

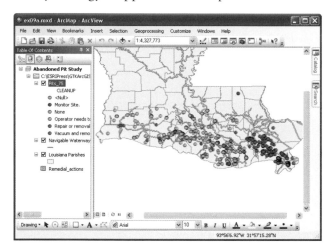

21 In the table of contents, place the mouse pointer over labels to see their full descriptions.

Pits that need to be repaired, removed, or treated are mostly located in a few parishes in the southeastern tip of the state.

22 If you want to save your work, click the File menu and click Save As. Navigate to **\GTKArcGIS\Chapter09\MyData**. Rename the file **my_ex09a.mxd** and click Save.

23 If you are continuing with the next exercise, leave ArcMap open. Otherwise, exit the application. Click No if prompted to save your changes.

Relating tables

Tables should be associated by a relate instead of a join when a record in the layer attribute table may have many matches in the nonspatial table (a one-to-many relationship). When tables are related, you can highlight records in either table to see matching records in the other.

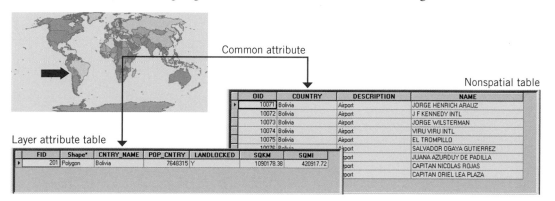

Common attribute

Nonspatial table

OID	COUNTRY	DESCRIPTION	NAME
10071	Bolivia	Airport	JORGE HENRICH ARAUZ
10072	Bolivia	Airport	J F KENNEDY INTL
10073	Bolivia	Airport	JORGE WILSTERMAN
10074	Bolivia	Airport	VIRU VIRU INTL
10075	Bolivia	Airport	EL TROMPILLO
10076	Bolivia	irport	SALVADOR OGAYA GUTIERREZ
		rport	JUANA AZURDUY DE PADILLA
		rport	CAPITAN NICOLAS ROJAS
		rport	CAPITAN ORIEL LEA PLAZA

Layer attribute table

FID	Shape*	CNTRY_NAME	POP_CNTRY	LANDLOCKED	SQKM	SQMI
201	Polygon	Bolivia	7648315	Y	1090178.38	420917.72

A nonspatial table of world airports is related to a layer attribute table of countries. You can select a country on the map to see the names of all its airports. You can also select a record in the nonspatial table to highlight a feature on the map.

Exercise 9b

Oil leaks and spills are not the only dangers posed by abandoned pits. The pits often contain toxic metals—barium, lead, and zinc among others—that should not contaminate water. Soil and water samples taken during the pit study have been analyzed for the presence of these metals. In this exercise, you'll relate tables to see the metals analysis for pits within a half mile of the Mississippi River.

Besides examining individual pits for the metals they contain, it would be useful to have a series of maps showing the distribution of particular metals for all hazardous pits—a map of pits containing barium, another of pits containing chromium, and so on. You will not go that far in this exercise, but you will relate tables to select all pits that contain lead.

8

9

1   In ArcMap, open **ex09b.mxd** from the **C:\ESRIPress\GTKArcGIS\Chapter09** folder.

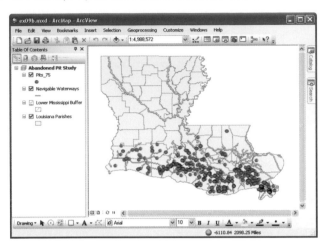

You see the map of Louisiana with abandoned pits, navigable waterways, and parishes. There is also a scale-dependent layer called Lower Mississippi Buffer.

2   Click the Bookmarks menu and click Lower Mississippi River.

The view zooms to a stretch of the Mississippi River. The Lower Mississippi Buffer layer is now visible, showing a half-mile buffer on either side of the river as it runs through southern Louisiana.

3   Click the Windows menu and click Overview. The Abandoned Pit Study Overview window opens and shows you the area you're zoomed to.

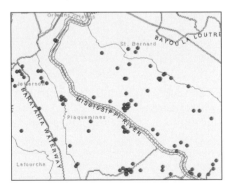 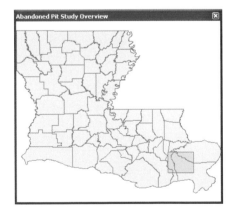

Of the pits in the Pits_75 layer, only two are within a half mile of the Mississippi. You'll check the toxic metals in each.

4 Close the overview window. Click the Bookmarks menu and click Site 1.

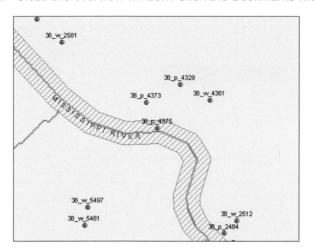

At this scale, the pits are labeled with their ID numbers.

5 On the Tools toolbar, click the Identify tool.

6 Click on pit 38_p_4375 to identify it.

The attributes of the Pits_75 layer do not include an analysis of metals found on-site. This information is stored in a separate table, which you'll add to the map document.

7 Close the Identify window. On the Standard toolbar, click the Add Data button.

8 In the Add Data dialog box, navigate to **\GTKArcGIS\Chapter09\Data**. Click on **Metals.dbf**, as shown in the following graphic, then click Add.

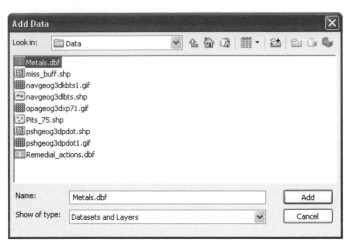

The Metals table is added to the table of contents and the List By Source tab is selected.

9 In the table of contents, right-click the Metals table and click Open.

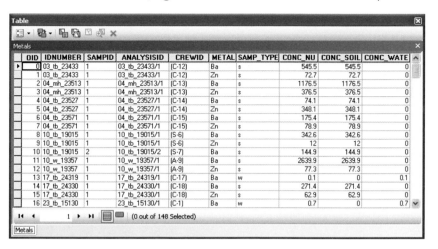

The table's attributes include METAL, which shows the chemical symbol for each metal sampled; SAMP_TYPE, either soil or water; and CONC_NUM, the concentration of the metal in the sample (in milligrams per kilogram for soil and milligrams per liter for water). The common attribute for the Metals and Pits_75 tables is IDNUMBER.

The Metals table has many records with the same ID number because there is a unique record for every sample taken at a pit. At most pits, several metals have been analyzed, and sometimes more than one sample of a metal has been taken.

Because the Pits_75 table contains one record per pit and the Metals table may contain several, the Pits_75 table has a one-to-many relationship to the Metals table. If you joined the tables, ArcMap would find the first matching record in the Metals table, join its attributes to the Pits_75 table, and ignore any further matching records. To preserve the one-to-many relationship, you must relate the tables instead.

10 Close the Attributes of Metals table. In the table of contents, right-click the Pits_75 layer, point to Joins and Relates, and click Relate.

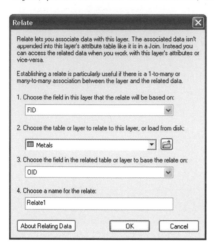

11 In drop-down list number 1 of the Relate dialog box, click the drop-down arrow and click IDNUMBER. This specifies IDNUMBER as the common attribute in the Pits_75 table.

12 In drop-down list number 2, make sure the Metals table is chosen. In number 3, make sure the IDNUMBER field is chosen. This specifies IDNUMBER as the common attribute in the Metals table.

13 In drop-down list number 4, highlight the default name and type **Metals Analysis**. Make sure your dialog box matches the following graphic, then click OK.

A relate is established between the two tables.

14 In the table of contents, click the List By Selection button. Under Selectable, click the toggle for the Navigable Waterways, Lower Mississippi Buffer and Louisiana Parishes layers so they are moved to Non Selectable. Only Pits_75 should be Selectable, as shown in the following graphic.

15 On the Tools toolbar, click the Select Features tool.

16 On the map, click pit 38_p_4375 to select it.

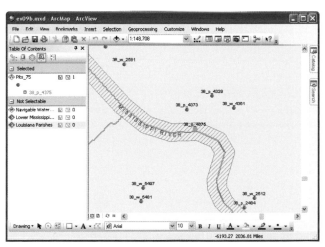

17 In the table of contents, right-click the Pits_75 layer and click Open Attribute Table.

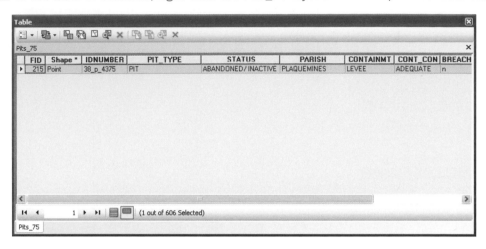

18 At the top left, click the Table Options menu. On the context menu, point to Related Tables and click Metals Analysis : Metals.

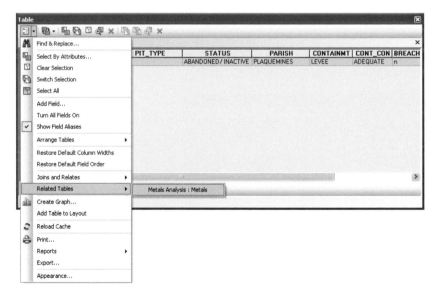

19 The Metals attribute table opens in a new tab. (Notice at the bottom of the table that the Pits_75 table is still available in a tab.)

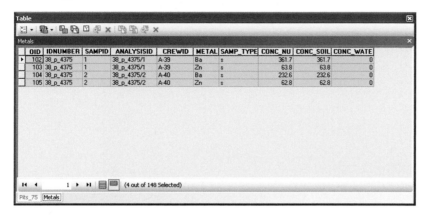

Four records in the Metals table match the record for Pit 38_p_4375. Two soil samples (SAMPID values 1 and 2) were taken from the pit and each sample was analyzed for barium and zinc (Ba and Zn in the METAL field). The concentration levels (CONC_NUM) are relatively low—barium samples for all pits have a mean value of 5,116, while zinc samples have a mean value of 410.

Now you'll check the levels at the other pit.

20 Close the attribute tables. Click the Bookmarks menu and click Site 2.

21 Make sure the Select Features tool is still selected on the Tools toolbar and click pit 38_p_4565 (the one in the buffer zone).

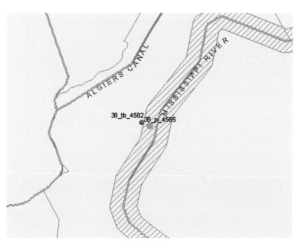

22 In the table of contents, right-click the Pits_75 layer and click Open Attribute table.

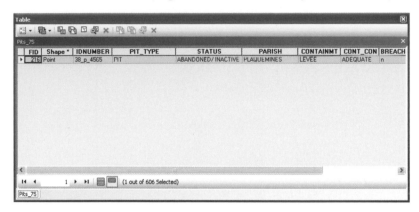

23 Click the Table Options menu, point to Related Tables, and click Metals Analysis : Metals.

24 The Metals attribute table opens in a new tab.

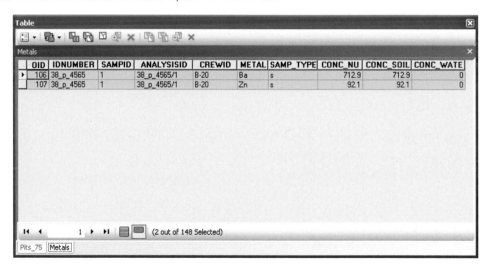

The pit has two matching records in the Metals table—a soil sample of barium and a soil sample of zinc. Again, both concentrations are relatively low.

Your final task is to locate all pits that contain lead. To do this, you must use the relate in the other direction: first making a selection on the Metals table and then seeing which features in the Pits_75 layer are selected.

25 At the bottom of the Metals table, click the Show all records button. Click the Table Options menu. On the context menu, click Select By Attributes.

26 Make sure the Method is set to Create a new selection. In the Fields scrolling box, double-click "METAL" to start the expression. Click the equals (=) button. Click Get Unique Values, then double-click "Pb," the symbol for lead. Make sure your expression matches the following graphic, then click Apply.

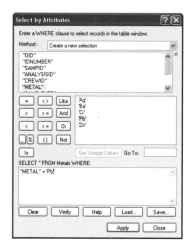

27 In the Metals attribute table, click the Show selected records button to show only the selected records.

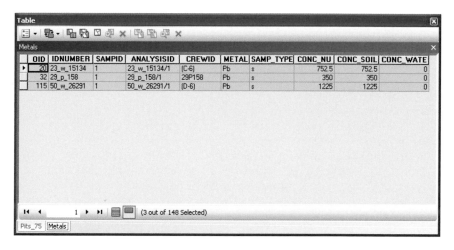

There are three records for lead samples. By comparing the ID numbers, you can see that each lead sample was taken from a different pit.

28 In the Metals attribute table, click the Table Options menu, point to Related Tables, and click Metals Analysis : Pits_75.

The Selected Attributes of Pits_75 table updates. Click its title bar to bring it forward.

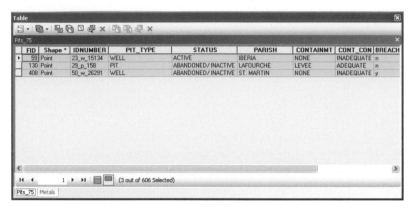

You see the records for the three pits that contain lead.

29 Close the attribute tables. (The Select By Attributes dialog box closes by itself.) In the table of contents, right-click the Pits_75 layer, point to Selection, and click Zoom To Selected Features.

ArcMap zooms to the smallest extent that shows all selected features.

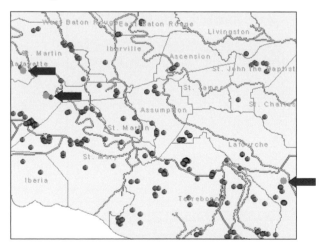

To make a map of pits containing lead, you would create a layer from the selected features. You'll learn how to create a selection layer in chapter 11.

An interesting follow-up to your work would be to join the Remedial_actions table to the Pits_75 table, then relate the Metals table to the joined table. You could then see which cleanup options have been recommended for the three sites that contain lead.

30 If you want to save your work, save it as **my_ex09b.mxd** in the **\GTKArcGIS\Chapter09 \MyData** folder.

When you save a map document, joins and relates are saved with it and are restored the next time you open the document. If you want to remove a join, right-click the layer with joined attributes, point to Joins and Relates, point to Remove Join(s), and click the name of the join. To remove a relate, point to Remove Relate(s) and click the name of the relate.

31 If you are continuing to the next chapter, leave ArcMap open. Otherwise, exit the application. Click No when prompted to save your changes.

Chapter 10

Selecting features by location

Using location queries

Combining attribute and location queries

In chapters 8 and 9, you selected features according to their attribute values. In this chapter, you'll select features by location—that is, according to their spatial relationship to other features, whether in another layer or in the same layer.

The four types of spatial relationships that can be used are distance, containment, intersection, and adjacency.

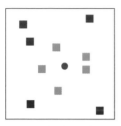

Points within a given distance of the red point are selected.

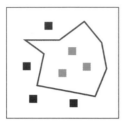

Points contained by the red polygon are selected.

Lines that intersect the red line are selected.

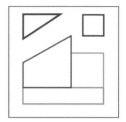

Polygons adjacent to the red polygon are selected.

ArcGIS defines thirteen spatial relationships, each a variation on one of the four types. For example, there is containment (where the contained feature may touch the boundary of the containing feature) and complete containment (where the boundaries may not touch).

Using location queries

To select features by location, you choose two layers and specify a spatial relationship. Features in the first layer are then spatially compared to features in the second layer and selected if they satisfy the relationship.

If the spatial relationship is one of distance, you also choose the distance value and measurement units you want to apply. In the following example, cities are selected if they lie within one mile of rivers.

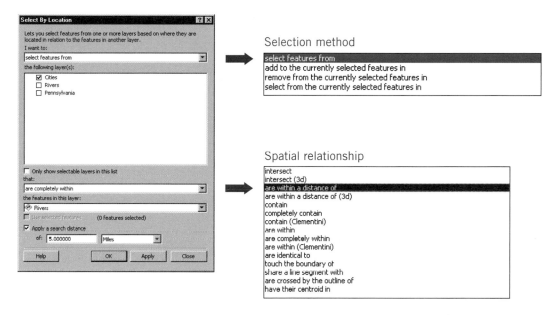

ArcMap measures the distances from cities to rivers and selects cities that satisfy the relationship.

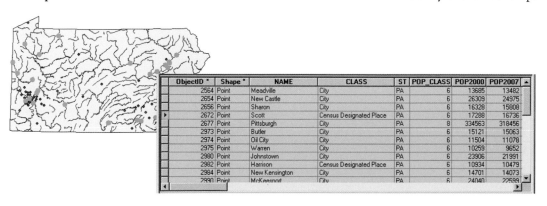

For more information, click the Contents tab in ArcGIS Desktop Help and navigate to *Professional Library > Mapping and Visualization > Working with layers > Interacting with layer contents > Using Select by Location.*

Exercise 10a

You work for a small chain of gourmet food stores that is scouting locations for a new store in Riverside, California.

The location should be close to a freeway so it is accessible to shoppers in nearby cities and people coming home from work. It should be in or near a shopping center to be convenient for people running other errands. And because your products are expensive, the store should be in an area with a large number of affluent households.

You'll begin by selecting neighborhoods that contain shopping centers and are close to freeways.

Some data in this chapter has been fictionalized to fit the scenario.

1   Start ArcMap. In the ArcMap—Getting Started dialog box, under the Existing Maps section, click Browse for more. (If ArcMap is already running, click the File menu and click Open.) Navigate to **C:\ESRIPress\GTKArcGIS\Chapter10**. Click **ex10a.mxd** and click Open.

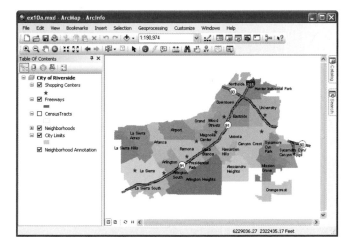

The map extent is the city limits of Riverside. Layers include neighborhoods, freeways, and shopping centers.

First, you'll select neighborhoods near freeways.

2   Click the Selection menu and click Select By Location.

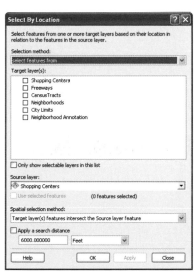

By default, the selection method is "select features from." Use this whenever you want to create a new selected set.

3 In the box of layers, check Neighborhoods.

4 In the list of Source layers, click the drop-down arrow and click Freeways.

5 In the list of Spatial selection methods, click the drop-down arrow and click "Target layer(s) features are within a distance of the Source layer feature."

The check box to apply a search distance to the features is checked. This search distance (not drawn on the map) defines the distance around freeways within which features will be selected.

6 Replace the current distance value with **0.5**. Click the distance units drop-down arrow and click Miles.

You have specified that you want to select all neighborhoods within half a mile of a freeway. A neighborhood will be selected if any part of it is within this distance.

7 Make sure your dialog box matches the following graphic, then click Apply and move it away from the map.

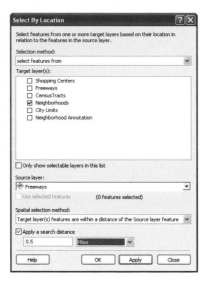

The neighborhoods are selected on the map.

Now you'll refine the selection by choosing those neighborhoods from the currently selected set that also contain a shopping center.

8 In the Select By Location dialog box, click the selection method drop-down arrow (the one at the top) and click "select from the currently selected features in." In the layers box, leave Neighborhoods checked.

9 In the drop-down list of layers, click Shopping Centers.

10 In the list of Spatial selection methods, click the drop-down arrow and click "Target layer(s) features contains the Source layer feature." Leave the check box Apply for search distance unchecked.

ArcMap will examine the selected neighborhoods to see which ones contain a shopping center. Those that do will remain selected; those that don't will be unselected.

11 Make sure your dialog box matches the following graphic. Click Apply, then click Close.

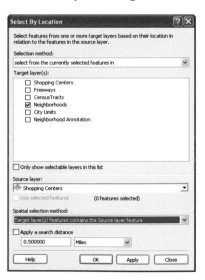

10
11
12
13

The neighborhoods are selected on the map.

Six neighborhoods meet your criteria. In the next exercise, you'll narrow the search further by looking for the affluent ones.

12 If you want to save your work, click the File menu and click Save As. Navigate to **\GTKArcGIS\Chapter10\MyData**. Rename the file **my_ex10a.mxd** and click Save.

13 If you are continuing with the next exercise, leave ArcMap open. Otherwise, exit the application. Click No if prompted to save your changes.

Combining attribute and location queries

Location and attribute queries can be used together to solve a problem. For example, if you're doing emergency planning, you might want to select cities of 100,000 or more (an attribute query) that lie within 10 miles of major fault lines (a location query).

Exercise 10b

You have found neighborhoods that are close to freeways and contain shopping centers. You also want to locate your store in an affluent neighborhood. The Neighborhoods layer has no demographic information, but the Census Tracts layer does. You'll do an attribute query to select the census tracts that meet your criteria. Then you'll create a location query to select the neighborhoods that substantially overlap these census tracts. Once you decide on a neighborhood, you'll add zoning and building data to help you choose a specific location for the new store.

1   In ArcMap, open **ex10b.mxd** from the **C:\ESRIPress\GTKArcGIS\Chapter10** folder.

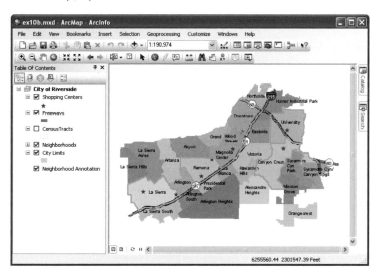

The map displays the city of Riverside and the neighborhoods selected in the previous exercise.

10

11

12

13

2   In the table of contents, turn on the Census Tracts layer.

Thin light lines outline the census tracts. Some tracts cross neighborhood boundaries and some neighborhoods include more than one tract.

3   In the table of contents, right-click the Census Tracts layer and click Open Attribute Table.

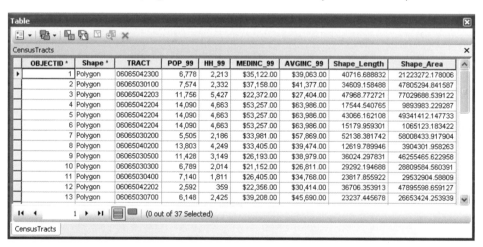

OBJECTID *	Shape *	TRACT	POP_99	HH_99	MEDINC_99	AVGINC_99	Shape_Length	Shape_Area
1	Polygon	06065042300	6,778	2,213	$35,122.00	$39,063.00	40716.688832	21223272.178006
2	Polygon	06065030100	7,574	2,332	$37,158.00	$41,377.00	34609.158488	47805294.841587
3	Polygon	06065042203	11,756	5,427	$22,372.00	$27,404.00	47968.772721	77029688.539122
4	Polygon	06065042204	14,090	4,663	$53,257.00	$63,986.00	17544.540765	9893983.229287
5	Polygon	06065042204	14,090	4,663	$53,257.00	$63,986.00	43066.162108	49341412.147733
6	Polygon	06065042204	14,090	4,663	$53,257.00	$63,986.00	15179.959301	1065123.183422
7	Polygon	06065030200	5,505	2,186	$33,981.00	$57,869.00	52138.381742	58008433.917904
8	Polygon	06065040200	13,803	4,249	$33,405.00	$39,474.00	12619.789946	3904301.958263
9	Polygon	06065030500	11,428	3,149	$26,193.00	$38,979.00	36024.297831	46255465.622958
10	Polygon	06065030300	6,789	2,014	$21,152.00	$26,811.00	29292.194688	28809584.560391
11	Polygon	06065030400	7,140	1,811	$26,405.00	$34,768.00	23817.855922	29532904.58809
12	Polygon	06065042202	2,592	359	$22,356.00	$30,414.00	36706.353913	47895598.659127
13	Polygon	06065030700	6,148	2,425	$39,208.00	$45,690.00	23237.445678	26653424.253939

You'll use the number of households (HH_99) and the average household income (AVGINC_99) as criteria for locating your business.

4 At the top of the table, click the Table Options menu and click Select By Attributes.

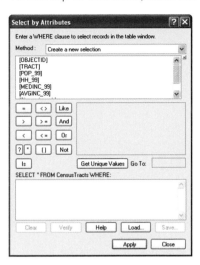

The Method drop-down list is set to Create a new selection, which is what you want.

5 In the Fields box, double-click [HH_99]. Click the > (Greater Than) button. Press the space bar and type **4000**. Click the And button. In the Fields box, double-click [AVGINC_99]. Click the > button. Press the space bar and type **65000**.

Your expression will select census tracts with more than four thousand households and where the average yearly income is over $65,000.

6 Make sure your dialog box matches the following graphic, then click Apply. Close the attribute table. The Select by Attributes dialog box closes with it.

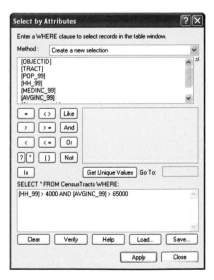

In the display, three census tracts are selected.

It's hard to pick out the selected tracts because they are highlighted in the same color as the neighborhoods.

7   In the table of contents, turn off the Neighborhoods layer.

The three tracts are located in the southeastern part of the city.

8   In the table of contents, turn the Neighborhoods layer back on.

You'll use the selected census tracts to select the neighborhoods they overlap.

9 Click the Selection menu and click Select By Location.

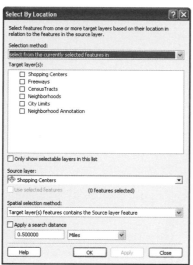

10 If necessary, click the selection method drop-down arrow and click "select from the currently selected features in."

11 In the Target layer(s) box, check the Neighborhoods check box.

12 In the list of Source layers, click the drop-down arrow and click Census Tracts.

13 In the list of Spatial selection methods, click the drop-down arrow and click "Target layer(s) features have their centroid in the Source layer feature."

Neighborhoods with centers in one of the selected census tracts will be selected. This method will find those neighborhoods that substantially overlap the selected census tracts (and therefore substantially share their demographic characteristics).

Underneath the Source layer drop-down list, the check box to use selected features is checked. This means that only the three selected features in the Census Tracts layer will be used in the spatial selection.

14 Make sure your dialog box matches the following graphic, then click Apply. When the selection is finished, click Close.

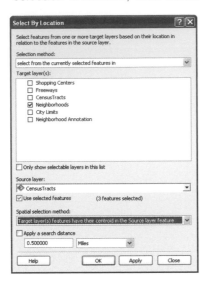

15 In the table of contents, turn off the Census Tracts layer.

A single neighborhood, Canyon Crest, is selected. It is populous, affluent, contains a major shopping center, and is close to a freeway.

16 In the table of contents, right-click the Neighborhoods layer, point to Selection, and click Zoom To Selected Features.

The map zooms to the Canyon Crest neighborhood. To look for sites in this neighborhood, you'll add two new layers to the map. One shows which parts of the neighborhood are zoned for commercial use. The other shows buildings within those districts.

17 On the Standard toolbar, click the Add Data button.

18 In the Add Data dialog box, navigate to **\GTKArcGIS\Chapter10\Data**. Double-click **RiversideCityData.mdb** to see its contents. Click **CCbuildings**. Press the Ctrl key and click **CCzoning**. Make sure your dialog box matches the following graphic, then click Add.

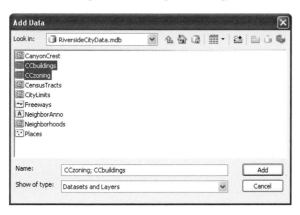

10
11
12
13

The two new layers are added to the map. (Your colors may be different.) The layers contain zoning and building data clipped to the Canyon Crest neighborhood. You'll learn how to clip data in the next chapter.

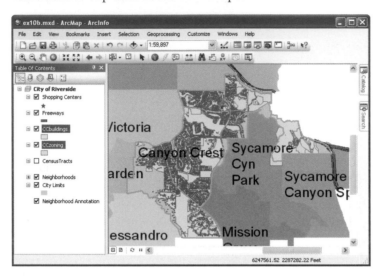

19 In the table of contents, turn off the CCbuildings layer, then double-click the CCzoning layer to open its layer properties. Click the Definition Query tab.

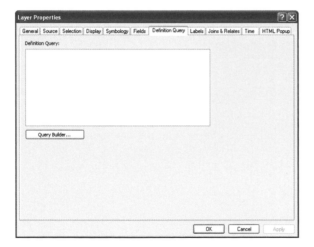

A definition query resembles an attribute query in that you write an expression to find features with particular attributes. The difference is that features satisfying an attribute query are selected, while features satisfying a definition query are displayed and the rest are hidden.

Since you're interested in looking only at commercial property, you'll write a query to display only those features in the CCzoning layer that are zoned for commercial use.

20 On the Definition Query tab, click Query Builder.

21 In the Fields box, double-click [DESC_] to add it to the expression box. (This attribute contains zoning code descriptions.) Click the equals (=) button. Click Get Unique Values, then double-click 'Commercial'. Make sure that your dialog box matches the following graphic, then click OK.

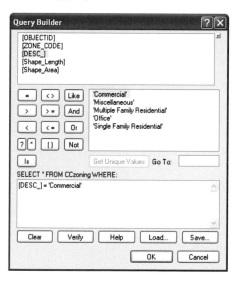

10

11

12

13

The query is displayed in the Definition Query box.

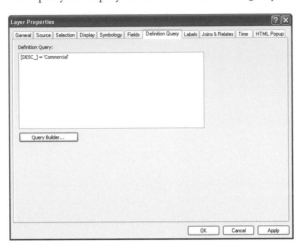

22 Click OK on the Layer Properties dialog box.

Now the only features displayed in the layer are those zoned for commercial use. Of the two commercial areas, shown by arrows in the previous graphic, you want the one that contains the shopping center.

23 On the Tools toolbar, click the Zoom In tool.

24 Drag a zoom rectangle around the commercial area containing the shopping center.

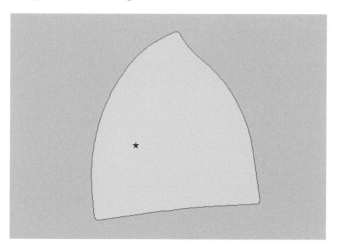

25 In the table of contents, turn on the CCbuildings layer.

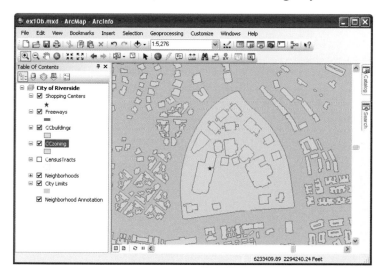

The commercially zoned buildings vary in size. You need 4,000 to 6,000 square feet for your store. You can identify the buildings and get their square footage from the Shape_ Area field in the layer attribute table. Since you don't know what the buildings look like or whether they are available, your next step might be to drive out to the area and then call a real estate agent.

26 If you want to save your work, save it as **my_ex10b.mxd** in the **\GTKArcGIS\Chapter10 \MyData** folder.

27 If you are continuing to the next chapter, leave ArcMap open. Otherwise, exit the application. Click No when prompted to save your changes.

Joining attributes by location

In this chapter, you selected features in one layer according to their spatial relationships to features in other layers. You can use the same kinds of spatial relationships (containment, distance, intersection) to join the attributes of features in one layer to features in another. This operation, called a spatial join, may also create new attributes, such as a count of the features in one layer that are contained by features in another layer or distance measurements between features in two layers.

Here, a layer of earthquake points is spatially joined to a layer of California counties. The output gives county-by-county totals for such earthquake attributes as damage and deaths.

Target layer (Counties)

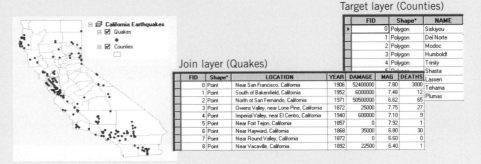

Join layer (Quakes)

FID	Shape*	LOCATION	YEAR	DAMAGE	MAG	DEATHS
0	Point	Near San Francisco, California	1906	52400000	7.80	3000
1	Point	South of Bakersfield, California	1952	6000000	7.48	12
2	Point	North of San Fernando, California	1971	50500000	6.62	65
3	Point	Owens Valley, near Lone Pine, California	1872	25000	7.75	27
4	Point	Imperial Valley, near El Centro, California	1940	600000	7.10	9
5	Point	Near Fort Tejon, California	1857	0	7.92	1
6	Point	Near Hayward, California	1868	35000	6.80	30
7	Point	Near Round Valley, California	1872	0	6.60	0
8	Point	Near Vacaville, California	1892	22500	6.40	1

FID	Shape*	NAME	Count_	Sum_DAMAGE	Sum_DEATHS
50	Polygon	San Luis Obispo	4	0	1
51	Polygon	Ventura	1	100000	0
52	Polygon	Riverside	8	625000	1
53	Polygon	Santa Barbara	6	0	0
54	Polygon	Orange	1	0	0
55	Polygon	San Diego	2	0	0
56	Polygon	Los Angeles	15	2091040000	176
57	Polygon	Imperial	15	2190000	24

This spatial join is based on containment. The output table has the attributes of the Counties layer plus three new ones. For each county, you can see how many earthquakes there have been, how much damage has been done, and how many people have died. In the map at the right, the counties are symbolized by number of earthquakes.

(Continued on next page)

Joining attributes by location (continued)

Here, the same layer of earthquake points is spatially joined to a layer of cities to find the distance from each city to the earthquake nearest it.

Target layer (Cities)

FID	Shape*	CITY_NAME
0	Point	McKinleyville
1	Point	Arcata
2	Point	Eureka
3	Point	Redding
4	Point	Red Bluff
		Paradise
		Chico
		Oroville
		Marysville

Join layer (Quakes)

FID	Shape*	LOCATION	YEAR	DAMAGE	MAG	DEATHS
0	Point	Near San Francisco, California	1906	52400000	7.80	3000
1	Point	South of Bakersfield, California	1952	6000000	7.48	12
2	Point	North of San Fernando, California	1971	50500000	6.62	65
3	Point	Owens Valley, near Lone Pine, California	1872	25000	7.75	27
4	Point	Imperial Valley, near El Centro, California	1940	600000	7.10	9
5	Point	Near Fort Tejon, California	1857	0	7.92	1
6	Point	Near Hayward, California	1868	35000	6.80	30
7	Point	Near Round Valley, California	1872	0	6.60	0
8	Point	Near Vacaville, California	1892	22500	6.40	1

This spatial join is based on distance. The output table has the attributes of both the Cities layer and the Quakes layer. The LOCATION attribute from the Quakes layer shows the nearest earthquake to each city. The new Distance attribute shows how far away the quake was (in meters).

Output layer (Counties)

FID	Shape	CITY_NAME	LOCATION	YEAR	DAMAGE	MAG	DEATHS	Distance
0	Point	San Francisco	Near San Francisco, California	1906	52400000	7.8	3000	12267.674055
1	Point	Daly City	Near San Francisco, California	1906	52400000	7.8	3000	2158.024984
2	Point	South San Francisco	Near San Francisco, California	1906	52400000	7.8	3000	7187.750121
3	Point	Lancaster	North of San Fernando, California	1971	50500000	6.62	65	36739.752995
4	Point	Palmdale	North of San Fernando, California	1971	50500000	6.62	65	34705.523132
5	Point	Santa Clarita	North of San Fernando, California	1971	50500000	6.62	65	9598.026073
6	Point	Calexico	Imperial Valley, near El Centro, California	1940	600000	7.1	9	6047.696612
7	Point	Castro Valley	Near Hayward, California	1868	35000	6.8	30	3452.466603
8	Point	San Leandro	Near Hayward, California	1868	35000	6.8	30	5076.885168
9	Point	Ashland	Near Hayward, California	1868	35000	6.8	30	1408.518612
10	Point	Cherryland	Near Hayward, California	1868	35000	6.8	30	2299.832637
11	Point	San Lorenzo	Near Hayward, California	1868	35000	6.8	30	5731.422385

10
11
12
13

You do a spatial join in much the same way as the table join you did in chapter 9. In the table of contents, right-click the target layer (the one you are joining attributes to), point to Joins and Relates, and click Join. In the Join Data dialog box, choose "Join data from another layer based on spatial location," then select the join layer. The output layer attributes vary according to the spatial relationship between the target and join layers and the options you choose in the dialog box. For more information, click the Contents tab in ArcGIS Desktop Help and navigate to *Professional Library > Data Management > Geographic data types > Tables > Joining tables by spatial query > About joining the attributes of features by their location*.

Chapter 11

Preparing data for analysis

Dissolving features
Creating graphs
Clipping layers
Exporting data

GIS analysis projects usually begin with several datasets and the plans to process them to get a result. When you perform an operation on spatial data that creates a new dataset (often a slightly modified version of the original), you are doing geoprocessing. Geoprocessing tasks are accomplished with Toolboxes, a collection of tools that you add to the ArcMap or ArcCatalog interface from the Standard toolbar.

Datasets are not always in exactly the condition you need for a project. For example, features may be either too detailed or too generalized for the map scale you are working at. Another problem may be not having all the attributes you want. And sometimes the problem is even one of too much data—you may have thousands of features cluttering your area of interest or extending well beyond it.

Having too much data is, of course, a better problem than not having enough. Toolboxes contain a number of tools to help you streamline datasets. You can simplify data by dissolving a group of features with a common attribute value into one feature. You can trim a dataset to your area of interest by using features in one layer to clip features in another. Other techniques don't require special geoprocessing tools. For example, you can work with fewer features by making a selection on a layer and creating a new layer from it.

Dissolving features

A dissolve creates a new layer in which all features in an input layer that have the same value for a specified attribute become a single feature. In the following example, states are dissolved by sales region.

States labeled by sales region

States dissolved by sales region

The new layer's attribute table has the standard geometry (Shape) and feature identifier (FID) attributes along with the attribute used in the dissolve (Region). You can include other attributes as well. In the sales region example, the input table has an attribute storing the number of customers per state. In the output table, these values can be summed for each region.

10
11
12
13

Attributes of States

FID*	Shape*	StateName	Customers	Region
42	Polygon	Alabama	52	South
36	Polygon	Arizona	31	West
46	Polygon	Arkansas	26	Central
24	Polygon	California	30	West
31	Polygon	Colorado	24	West
18	Polygon	Connecticut	24	East
28	Polygon	Delaware	23	East
27	Polygon	District of Columbia	22	East
48	Polygon	Florida	24	South

Record: 0   Show: All   Selected   Records (0 out of 49 Selected.)

Input sales region table

Attributes of Sales_Regions

FID*	Shape*	Region	SUM_Customers
1	Polygon	Central	295
2	Polygon	East	223
3	Polygon	South	233
4	Polygon	West	214

Record: 1   Show: All   Selected   Records (0 out of

Output sales region table

Exercise 11a

You work for a lumber company that plans to harvest timber in the Tongass National Forest in southeastern Alaska. Forest land can be divided into stands—groups of trees with something in common, such as type, age, or size. In a national forest, stands can be grouped into larger areas and leased to private companies by the U.S. Forest Service. Restrictions are placed on logging in sensitive parts of the lease areas, such as endangered animal habitat.

The Forest Service is presently considering leasing five adjacent areas.

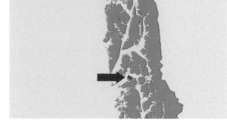

The Tongass National Forest, shown in dark green, covers 16,800,000 acres (about 68,000 square kilometers) of the Alaska panhandle.

Lease areas under consideration are shown in red.

As your company's GIS analyst, your job in this and the following chapter is to calculate the timber values of the potential lease areas. Your analysis will help your company decide how much to bid for each area.

You have a polygon layer of forest stands provided by the forest service. Its attributes include the estimated value of each stand and the lease area each stand belongs to.

In this exercise, you'll dissolve the stands into the five lease areas. You'll total the stand values to get a preliminary estimate of how much each lease area is worth. In chapter 12, you'll refine this estimate by eliminating areas that can't be harvested.

1   Start ArcMap. In the ArcMap—Getting Started dialog box, under the Existing Maps
    section, click Browse for more. (If ArcMap is already running, click the File menu and
    click Open.) Navigate to **C:\ESRIPress\GTKArcGIS\Chapter11**. Click **ex11a.mxd** and
    click Open.

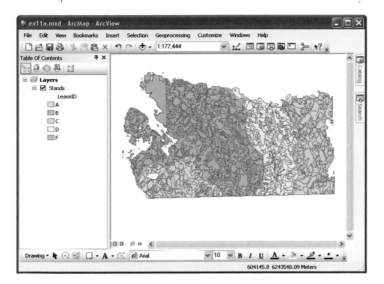

The map shows a layer of forest stands, symbolized by their respective lease areas.

2   In the table of contents, right-click the Stands layer and click Open Attribute Table.

OBJECTID	Shape *	LeaseID	StandValue	ValuePerMeter	StandID	Shape_Length	Shape_Area
1	Polygon	A	1.003531	64	6224	769.689245	15680.172005
2	Polygon	A	7.380323	42	1164	2164.996611	175721.964087
3	Polygon	A	0.936683	63	6223	650.045632	14867.984924
4	Polygon	A	10.568293	30	1169	5261.256657	352276.409512
5	Polygon	A	1.974708	63	1171	776.790314	31344.562658
6	Polygon	A	4.238298	63	1178	1307.944741	67274.578893
7	Polygon	A	5.056288	53	1198	1768.19406	95401.656306
8	Polygon	A	2.726222	52	1209	946.578676	52427.34269
9	Polygon	A	1.472038	53	1216	714.853307	27774.296188
10	Polygon	A	8.734157	25	1229	4421.968632	349366.263451
11	Polygon	A	1.022995	43	1233	687.696463	23790.5803
12	Polygon	A	7.126407	26	1259	3003.696897	274092.559487

The Stands layer has 1,405 stands, each represented by one record in the table.

The StandValue attribute contains the dollar value of each stand in millions of dollars. It
was obtained by multiplying ValuePerMeter by Shape_Area (Shape_Area stores the size
of each feature in square meters) and then dividing by 1,000,000. The value of the first
stand, for instance, appears as 1.00353, which is just over a million dollars.

3 Close the table. Click the Geoprocessing menu and click Search For Tools.

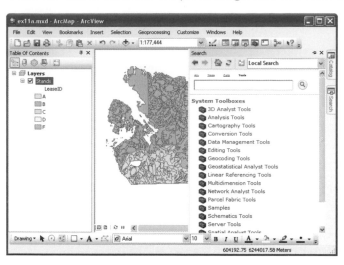

The Search window opens. If it is docked to the side of the main ArcMap window, it may look like the graphic above. (If you have opened it before and customized its position, your Search window may be in a different location.)

The Search window displays many system toolboxes. Each toolbox has tools for related geoprocessing tasks. You need the Dissolve tool, which is found in the Data Management Tools toolbox. (Alternatively, you could click the ArcToolbox window button, however, using the Search window is a convenient way to find a tool quickly.)

4 In the Search window, click the Data Management Tools link, then click the Generalization link. (You may need to scroll down to see it.)

The Search window graphics in this book assume that you are using an ArcView software license and that you have not installed optional extensions such as ArcGIS Spatial Analyst. If you have an ArcEditor or ArcInfo license, or if you have installed extensions, your window will show additional toolboxes.

5 Click the Dissolve tool to open the Dissolve dialog box.

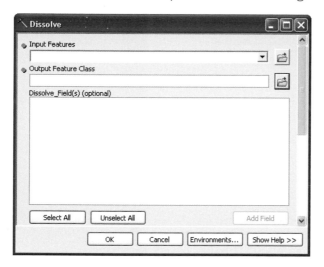

Although each tool has its own functionality and settings, all use a similar dialog box with some common elements. For example, each tool has a Show Help button in the lower right corner that describes the tool and its settings.

Tools with lots of settings have a scroll bar on the right. As an alternative to scrolling, you can resize the dialog box by dragging on a side or corner. Once you do this, ArcMap remembers the new size and applies it to the next tool you open. (So if the dimensions of your Dissolve dialog box are different from the graphic, it means you have previously resized this or some other tool. That's fine—the dialog box can be whatever size you like.)

The Dissolve tool requires that you select an input layer of features to dissolve, a location where the newly created layer will be saved to disk, and an attribute to dissolve on.

6 In the Input Features drop-down list, select Stands. (It's the only layer in the map.)

When you select the input features, a default path and file name appear for the output data. The list of attributes that can be dissolved on appears in the Dissolve_Field(s) box.

The output data can be saved as either a shapefile or as a geodatabase feature class. You'll save it as a geodatabase feature class because the rest of the Tongass data is in this format. To leave the Tongass geodatabase intact, you'll save the output to a duplicate geodatabase called MyTongass that has been created for you.

7 Click the Browse button next to the Output Feature Class box. In the Output Feature Class dialog box, navigate to **\GTKArcGIS\Chapter11\MyData**. Double-click **MyTongass.mdb**.

10

11

12

13

8 In the Name box, type **Leases**. Make sure your dialog box matches the following graphic, then click Save.

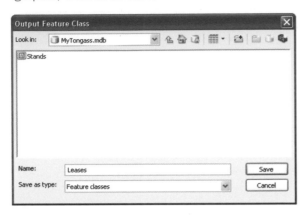

The output feature class information is updated.

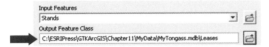

9 In the Dissolve_Field(s) box, click the LeaseID check box.

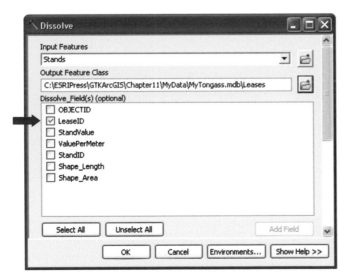

Each set of polygons with the same LeaseID will be dissolved into a single feature.

ArcMap can also summarize the attribute values of dissolved features by a variety of statistics. To find out the timber value of each lease, you will summarize the StandValue attribute by the Sum statistic type. This will total the values of all stands dissolved within a lease. (You can also get the mean, the range, the standard deviation, and other measures of numeric attributes.)

10 In the Dissolve dialog box, scroll down to the Statistics Field(s) area. Click the Statistics Field(s) drop-down arrow and click StandValue to add it to the list.

The field name displays in the Field column. The red circle with an x indicates that you still need to specify the type of statistic you want.

11 Click the first cell in the Statistic Type column. Click the drop-down list arrow and click SUM.

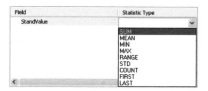

12 Make sure your Statistics Field(s) list matches the following graphic, then click OK.

10
11
12
13

Tools execute in the background, meaning you can continue working with ArcMap while the tool executes. You'll see a progress bar at the bottom of the main window displaying the name of the currently executing tool and the percentage to completion. When the tool finishes executing, a pop-up notification will appear on the system tray.

13 When the operation is completed, close the notification window (or the pop-up notification will disappear after a few seconds.)

The new Leases layer is added to the map. (If you don't see it, click the Add Data button, navigate to the MyTongass geodatabase, and add the layer. Then click the Geoprocessing menu and click Geoprocessing Options. In the Geoprocessing Options dialog box, near the bottom of the dialog box, check the box Add results of geoprocessing operations to the display box. Click OK. Any new geoprocessing layers will be added to ArcMap automatically.)

When a new geoprocessing layer is created and added to ArcMap, it displays in a random color. In many of the exercise steps in this chapter and the next, the colors on your screen may not match the graphics in the book.

14 In the table of contents, right-click the Leases layer and click Zoom to Layer.

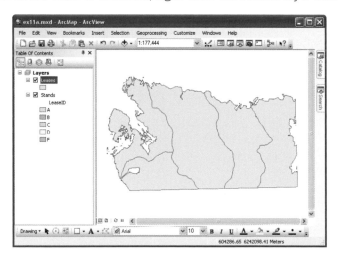

15 In the table of contents, right-click the Leases layer and click Open Attribute Table. If necessary, resize the table to show all fields.

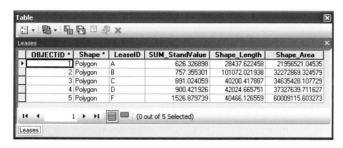

The original 1,405 stands have been aggregated into five lease polygons.

LeaseID is the attribute you dissolved on. SUM_StandValue is the statistical field you requested. It contains the sum of all stand values (in millions of dollars) in each lease area. The value of lease A, for instance, is about 626 million dollars.

Shape_Length and Shape_Area are measurement attributes automatically maintained by ArcMap for geodatabase feature classes. (Shape_Length, for a polygon feature class, measures feature perimeters.)

16 Close the table.

17 If you want to save your work, click the File menu and click Save As. Navigate to **\GTKArcGIS\Chapter11\MyData**. Rename the file **my_ex11a.mxd** and click Save.

18 If you are continuing with the next exercise, leave ArcMap open. Otherwise, exit the application. Click No if prompted to save your changes.

10

11

12

13

Creating graphs

The ArcGIS Graph Wizard lets you create many different kinds of graphs, including column, pie, area, and scatter graphs. You can set properties for such elements as titles, axes, and graph markers (the bars in a bar graph, for instance). Graphs can be saved with a map document or as files with a .grf extension that can be added to any map document.

Exercise 11b

You have dissolved the forest stands into lease areas and summed their harvestable values. In this exercise, you'll present the values in a graph and add the graph to a map layout.

1   In ArcMap, open **ex11b.mxd** from the **C:\ESRIPress\GTKArcGIS\Chapter11** folder.

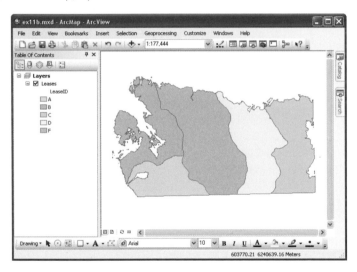

The map shows the lease areas you created in the previous exercise.

2 Click the Views menu, point to Graphs, and click Create.

The Create Graph Wizard opens.

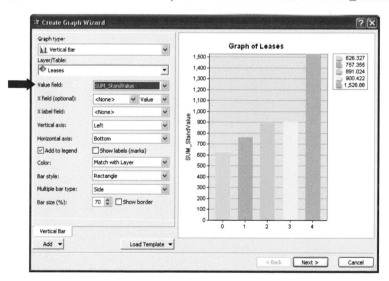

In the first panel, you'll accept the default graph type (Vertical Bar) and the default Layer (Leases), and choose the attribute you want to graph.

3 Click the Value field drop-down arrow and choose SUM_StandValue from the list.

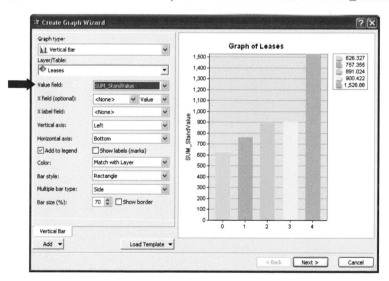

You'll label the x-axis. As you make changes, the graph will update.

10

11

12

13

4   Click the drop-down arrow for X label field and choose LeaseID from the list.

The LeaseID values display on the x-axis.

5   Uncheck Add to legend. Make sure your wizard matches the following graphic.

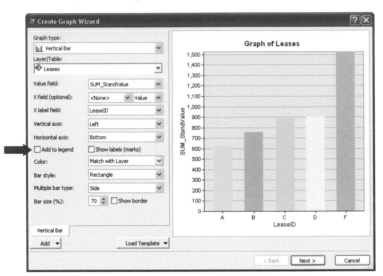

6   Click Next.

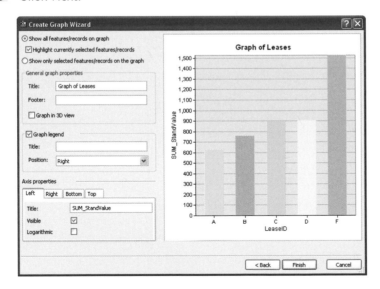

In the second panel, you'll change the default graph title, add a footer, and remove the axis titles.

7 In the Title box under General graph properties, replace Graph of Leases with **Lease Values**. In the Footer box, type **In millions of dollars**.

8 In the Axis properties panel (near the bottom of the wizard), make sure the Left tab is selected. Delete Sum_StandValue from the Title box, leaving it blank. Make sure Visible is checked.

Axis properties

Left	Right	Bottom	Top

Title:

Visible ☑

Logarithmic ☐

The left axis values display on the graph but the title "Sum_StandValue" does not. You'll make the same change to the bottom axis.

9 Make sure your wizard matches the following graphic, then click Finish.

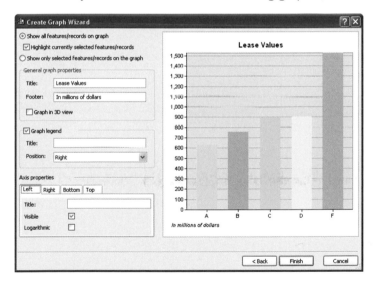

The graph displays in a window that floats on the application window. The graph makes it easy to compare the lease area values. Lease F, the most valuable, is worth about 1.5 billion dollars.

10
11
12
13

Note: If you resize the graph window, values displayed along the left axis may change.

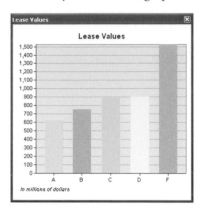

You'll add the graph to the map layout.

10   Right-click the graph title bar. On the context menu, click Add to Layout.

ArcMap switches to layout view.

11   Close the graph window.

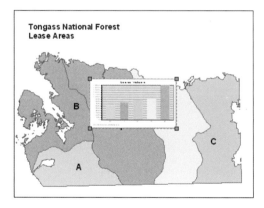

The graph displays in the middle of the layout page, marked with blue selection handles. (Other layout elements, such as the map title and lease labels, have been added for you.)

12   On the Tools toolbar, make sure the Select Elements tool is selected.

13 Drag the graph to the upper right corner of the layout, as shown in the following graphic. Click outside the layout page to unselect the graph.

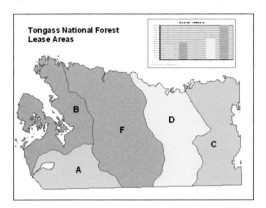

14 If you want to save your work, save it as **my_ex11b.mxd** in the **\GTKArcGIS\Chapter11 \MyData** folder.

15 If you are continuing with the next exercise, leave ArcMap open. Otherwise, exit the application. Click No if prompted to save your changes.

10

11

12

13

Clipping layers

Clipping trims features in one layer using the boundaries of polygon features from another layer. It's like having a pair of scissors, or a cookie cutter, to cut away data that you don't need for your project. For example, you might have a shapefile of all the streets in your county, while your study area encompasses just one ZIP Code within that county. If you have a polygon feature representing that ZIP Code, you can use it to clip out just the streets you need. Clipping can be a matter of convenience—since smaller datasets are easier to process—but it can also be important for analysis. To find out the total length of roadway within the ZIP Code, for instance, you must exclude streets and street segments that lie outside it.

In the following example, a layer of interstate highways is clipped at the boundaries of the state of Oklahoma.

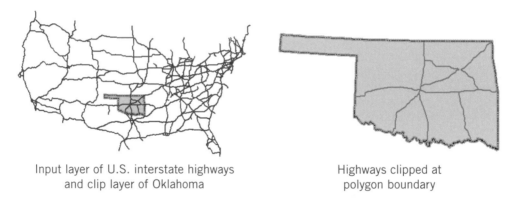

Input layer of U.S. interstate highways
and clip layer of Oklahoma

Highways clipped at
polygon boundary

Exercise 11c

The forest service has just determined that, of the five lease areas, only lease F is mature enough for harvest. Your company will direct its attention to making a bid for this lease.

Not every square meter of the lease area is harvestable. Logging is prohibited near streams and goshawk nests. (Goshawks are a protected bird species.) You have layers of streams and goshawk nests that cover all five lease areas, but now you'd like to work with datasets that cover only the area of lease F.

In this exercise, you'll clip a layer of streams to the boundary of lease F.

1 In ArcMap, open **ex11c.mxd** from the **C:\ESRIPress\GTKArcGIS\Chapter11** folder.

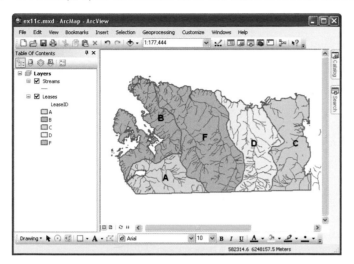

The map displays the five lease areas and a Streams layer that contains 1,566 stream segments.

To clip the streams to the lease F boundary, you must first select lease F. You'll turn the selection into a new layer so you can look at lease F apart from the others.

2 Click the List by Selection button. Click the toggle beside Streams to make it not selectable, as shown in the following graphic.

3 On the Tools toolbar, click the Select Features tool.

4 On the map, click lease F to select it. It is outlined in cyan.

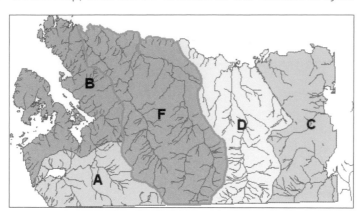

5 In the table of contents, right-click the Leases layer and click Create Layer From Selected Features.

A layer called Leases selection is added to the top of the table of contents. This layer references the same geodatabase feature class as the Leases layer. Unlike geoprocessing operations, creating a selection layer doesn't create a new dataset.

(You can, however, export a selection layer, or a set of selected features, as a new dataset if you want to. You will do this in the next exercise.)

6 Click the white area of the map to deselect the feature.

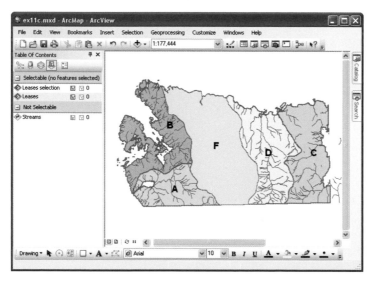

7 Click the List By Drawing Order button in the table of contents. Turn off the Leases layer and drag the Leases selection layer to the bottom of the table of contents.

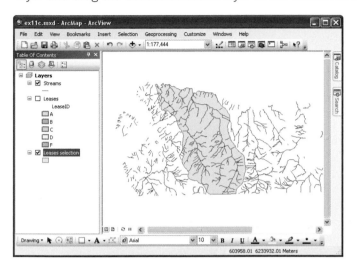

Commonly used geoprocessing tools are listed in the Geoprocessing menu. Alternatively, search for more tools in the ArcToolbox, Search or Catalog window. You'll use the Clip tool to clip the streams to the boundary of lease F.

8 Click the Geoprocessing menu and click Clip.

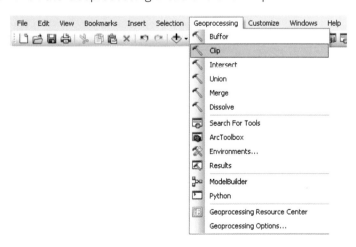

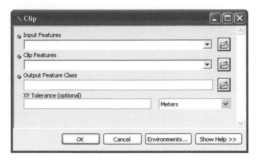

You'll select the layer with the input features to be clipped, the layer to clip with, and an output feature class. As with Dissolve, the output of a clip can be either a shapefile or a geodatabase feature class. You will add the clipped streams to the MyTongass geodatabase.

9 In the Input Features drop-down list, click Streams. In the Clip Features drop-down list, click Leases selection.

10 Click the Browse button next to the Output Feature Class drop-down list. In the Output Feature Class dialog box, navigate to **\GTKArcGIS\Chapter11\MyData** and double-click **MyTongass.mdb**.

11 In the Name box, type **StreamsF**. Make sure your dialog box matches the following graphic, then click Save.

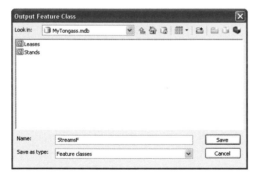

The output feature class information is updated in the Clip dialog box.

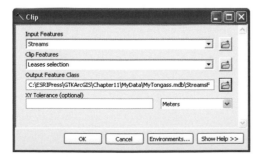

12 Click OK.

When the Clip tool finishes processing, a pop-up notification will appear on the system tray. Click the Clip link in the pop-up notification to view the Clip tool's results report. The Results window shows a summary of the input and output layers, environment parameters, each operation and the elapsed time.

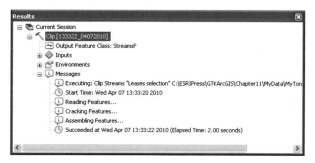

13 When the operation is completed, close the Results window. In the table of contents, turn off the Streams layer.

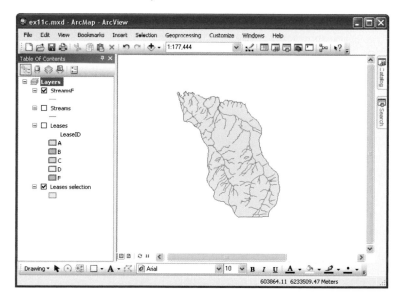

The StreamsF layer (assigned a random color), consisting of stream features clipped to the boundary of lease F, is added to the map.

10
11
12
13

14 In the table of contents, right-click the StreamsF layer and click Open Attribute Table.

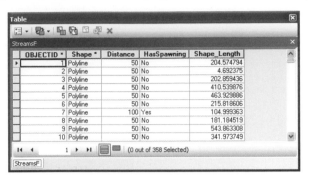

The Clip operation has reduced the number of stream segments from 1,566 (in the Streams layer) to 358. The table has attributes that will determine how much area around the streams is off-limits to logging. You will work with these attributes in the next chapter.

15 Close the table.

16 If you want to save your work, save it as **my_ex11c.mxd** in the **\GTKArcGIS\Chapter11 \MyData** folder.

17 If you are continuing with the next exercise, leave ArcMap open. Otherwise, close the application. Click No if prompted to save your changes.

Exporting data

Operations like Dissolve and Clip create new datasets automatically. Another way to make a new dataset from an existing one is to make a selection on a layer and export the selected features.

Exercise 11d

In addition to the streams layer, you have a layer of goshawk nests covering all five lease areas. You could clip this layer just as you did the streams. Unlike the streams, however, the nests are points and therefore don't cross polygon boundaries. (In the map it will look as if they do, but that is an effect of symbology.) Because nests are contained within polygons, you can use Select By Location to select the nests in lease F and then export the selected set.

1   In ArcMap, open **ex11d.mxd** from the **C:\ESRIPress\GTKArcGIS\Chapter11** folder.

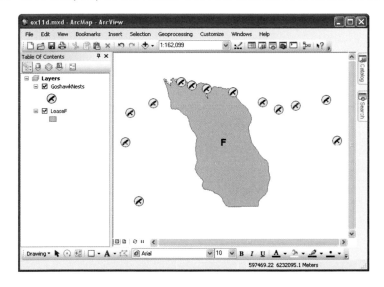

The map contains a layer of goshawk nests and a layer of lease F.

2   Click the Selection menu and click Select By Location. If necessary, click the selection method drop-down arrow and click "select features from."

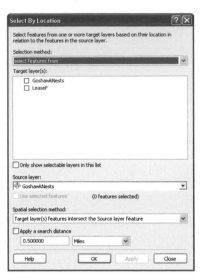

3   In the box of layers to select from, check GoshawkNests.

4   In the Source layer drop-down list, make sure LeaseF is chosen.

5   In the drop-down list of Spatial selection methods, click "Target layer(s) features are completely within the Source layer feature." Make sure your dialog box matches the following graphic, then click Apply.

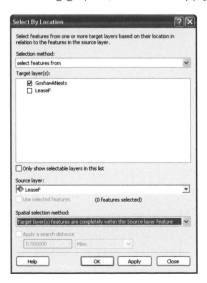

On the map, the four nests in lease F are selected.

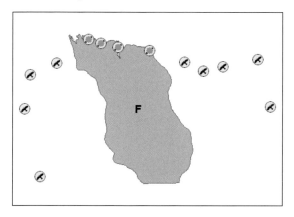

6 Close the Select By Location dialog box. In the ArcMap table of contents, right-click the GoshawkNests layer, point to Data, and click Export Data.

The Export drop-down list is correctly set to Selected features. The option to use the same coordinate system as the layer's source data is also correct. (You'll learn more about coordinate systems in chapter 13.)

7 Click the Browse button next to the Output shapefile or feature class box.

8 In the Saving Data dialog box, make sure the Save as type drop-down list is set to File and Personal Geodatabase feature classes. If necessary, navigate to **\GTKArcGIS \Chapter11\MyData** and double-click **MyTongass.mdb**.

9 In the Name box, replace Export_Output with **NestsF**. Make sure your dialog box matches the following graphic, then click Save.

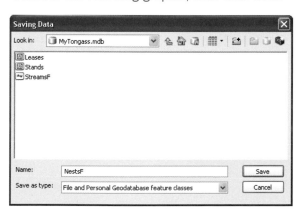

10 Click OK in the Export Data dialog box.

ArcMap exports the selected nests to a new feature class and prompts you to add the data to the map.

11 Click Yes.

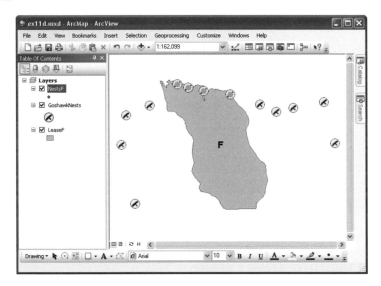

You don't see the point features of the new NestsF layer because they are covered by the selection highlights on the GoshawkNests layer.

12 In the table of contents, turn off the GoshawkNests layer.

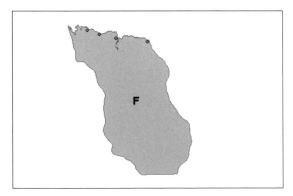

The new layer contains only goshawk nests within lease F. The default symbol is not as informative as the one in the GoshawkNests layer.

13 In the table of contents, double-click the NestsF layer. In the Layer Properties dialog box, click the Symbology tab and click Import.

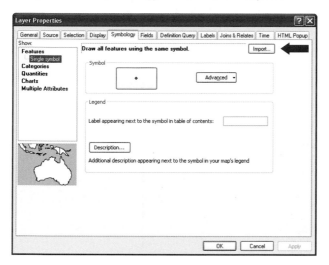

The Import Symbology dialog box opens.

You want to import symbology from another layer in the map, so the first option is set correctly. The Layer drop-down list is also correctly set.

14 Click OK.

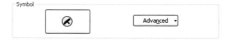

Notice that the symbology has changed.

15 Click OK in the Layer Properties dialog box to close it.

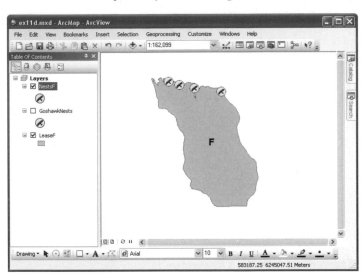

You have prepared the data for your analysis. In the next chapter, you'll create exclusion zones around the streams and goshawk nests in lease F. You'll use these zones to figure out how much timberland in the lease area is harvestable and how much it's worth.

16 If you want to save your work, save it as **my_ex11d.mxd** in the **\GTKArcGIS\Chapter11 \MyData** folder.

17 If you are continuing to the next chapter, leave ArcMap open. Otherwise, exit the application. Click No when prompted to save your changes.

Section 5: **Analyzing feature relationships**

Chapter 12

Analyzing spatial data

Buffering features
Overlaying data
Calculating attribute values

Most of the problems you solve with GIS involve comparing spatial relationships among features—in one layer or in different layers—and drawing conclusions. Problem solving in GIS is called spatial analysis, and it can include everything from measuring the distance between points to modeling the behavior of ecosystems.

The geoprocessing tools in ArcToolbox not only help you prepare data, they also help you analyze it spatially. In this chapter, you'll work with two tools that are very useful in spatial analysis: buffers and overlays.

A buffer is an area drawn at a uniform distance around a feature. It represents a critical zone, such as a floodplain, a protected species habitat, or a municipal service area. Features lying inside the buffer have a different status from features lying outside the buffer.

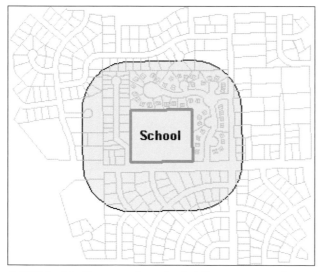

A 500-foot buffer around a school defines an area where billboard advertising is prohibited.

Overlays (union and intersect) identify overlaps between features in two layers, and create a dataset in which the lines of overlap define new features. In a union overlay, nonoverlapping areas are included in the output dataset. A union dataset, then, has three types of features: those found only in the first input layer (with layer 1 attribute values), those found only in the second input layer (with layer 2 attribute values), and those created by areas of overlap between the two layers (with both layer 1 and layer 2 attribute values).

In an intersect overlay, only the overlapping geometry is preserved, and features have attributes from both input layers.

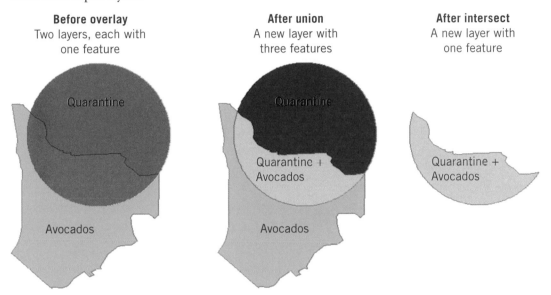

A fruit fly quarantine zone overlaps avocado groves (left). Overlay analysis—whether union or intersect—creates new features where input layers overlap. In this case, the analysis identifies areas where avocados must be destroyed.

Buffering features

Buffers are created as polygons in a new layer. Buffers can be drawn at a constant distance (for example, 100 meters) around every feature in a layer, or at a distance that varies according to attribute values. For example, buffers representing the range of radio signals from transmitters might vary according to an attribute describing the transmitter strength. Buffers can also be concentric rings representing multiple distances, such as the areas within 100, 500, and 1,000 meters of a well.

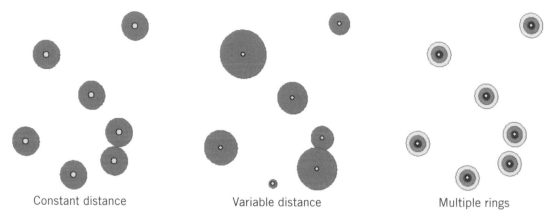

Constant distance   Variable distance   Multiple rings

If features are close together, their buffers may overlap. You can preserve the overlaps or remove them.

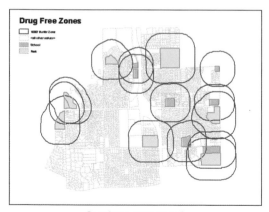

Overlaps preserved

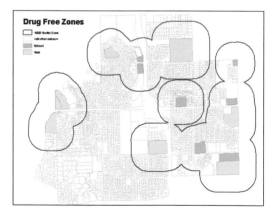

Multiple rings

Exercise 12a

Your goal is to determine the value of harvestable land in lease F so that your lumber company can make a bid. In chapter 11, you dissolved forest stands into leases. Then you clipped streams and selected goshawk nests within lease F. In this exercise, you'll buffer the nest and stream layers to show where logging is prohibited. According to government regulations, no trees may be cut within 800 meters of a goshawk nest, the range of goshawk fledglings. Nor can trees be cut within 50 meters of a stream. Logging near streams leads to erosion of the stream banks, adding sediment to the water. This kills aquatic plant life and disrupts the food chain. The prohibition on logging is increased to 100 meters from streams where salmon spawn.

1 Start ArcMap. In the ArcMap—Getting Started dialog box, under the Existing Maps section, click Browse for more. (If ArcMap is already running, click the File menu and click Open.) Navigate to **C:\ESRIPress\GTKArcGIS\Chapter12**. Click **ex12a.mxd** and click Open.

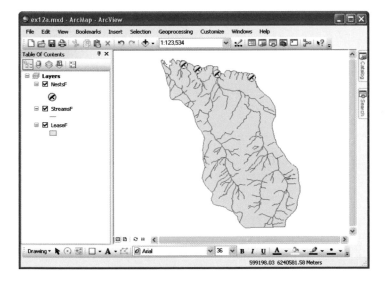

The map shows lease F, goshawk nests, and streams. You'll begin by buffering the goshawk nests using the ArcToolbox Buffer tool.

2 On the Standard toolbar, click ArcToolbox window button to open ArcToolbox.

3   In the ArcToolbox window, click the plus sign next to Analysis Tools. Click the plus sign next to Proximity.

Depending on how you set up your ArcToolbox window (floating, tabbed or docked), your display may look different. Also, depending on your ArcGIS license level, you may have more or fewer tools than the graphic above (which is using an ArcView license).

4   Double-click the Buffer tool.

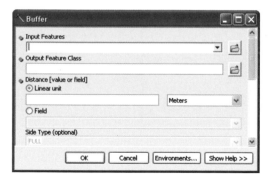

You'll select the layer with the input features to be buffered, designate an output feature class, enter a buffer distance, and select a dissolve option.

5   Click the Input Features drop-down arrow and click NestsF. (Alternatively, drag the NestsF layer from the ArcMap table of contents and drop it in this location.)

6   Click the Browse button next to the Output Feature Class box. In the Output Feature Class dialog box, navigate to **\GTKArcGIS\Chapter12\MyData** and double-click **MyTongass.mdb**.

7   In the Name box, type **NestBuf**. Make sure your dialog box matches the following graphic, then click Save.

The output feature class information is updated.

8   For the buffer distance, make sure the Linear unit option is selected. In the Linear unit box, type **800**.

The neighboring drop-down list shows the type of units, which are meters.

9   Scroll down if necessary, click the Dissolve Type drop-down arrow and click All.

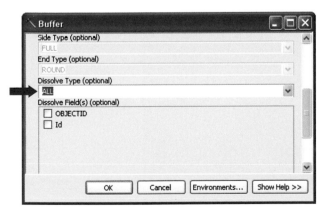

Wherever buffer polygons overlap each other, the overlapping boundaries will be dissolved to make a single feature.

10 Click OK.

When the Buffer tool finishes processing, a pop-up notification will appear on the system tray. Click the Buffer link in the pop-up notification to view the Buffer tool's results report.

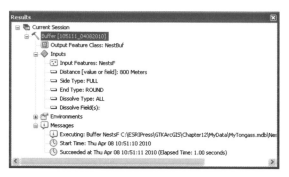

11 When the operation is completed, close the Results window.

The new NestBuf layer is added to the map. Where buffers overlap, the barriers between them have been removed, as you specified.

12 If necessary, change the color of the buffers so they are visible against the LeaseF background.

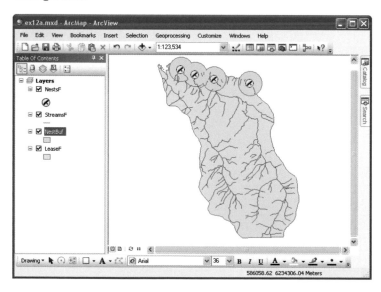

13 In the table of contents, right-click the NestBuf layer and click Open Attribute Table.

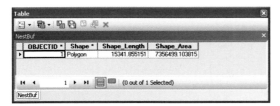

The table contains just a single record. Buffers created with the dissolve ALL option form a single feature called a multipart polygon (a polygon with discontinuous boundaries). The attributes are the standard four for a geodatabase polygon feature class: OBJECTID, Shape, Shape_Length, and Shape_Area. Because of the dissolve, none of the NestF layer attributes are passed on to the output table.

Now you'll buffer the Streams layer. The buffer sizes for this layer will vary according to whether or not salmon spawn in a stream.

14 Close the Attributes of NestBuf table. In the table of contents, right-click the StreamsF layer and click Open Attribute Table.

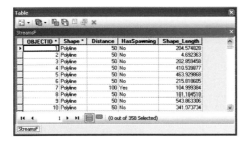

The HasSpawning field shows whether or not a stream has spawning salmon. The Distance field values of 50 and 100 correspond to the No and Yes values in the HasSpawning field. (Recall that logging is prohibited within 50 meters of streams, and within 100 meters of streams where salmon spawn.)

15 Close the attribute table. In the table of contents, turn off the NestBuf and NestsF layers.

16 In the ArcToolbox window, double-click the Buffer tool.

17 Click the Input Features drop-down arrow and click StreamsF.

18 Click the Browse button next to the Output Feature Class box. In the Output Feature Class dialog box, navigate to **\GTKArcGIS\Chapter12\MyData** and double-click **MyTongass.mdb**.

19 In the Name box, type **StreamsBuf**. Make sure your dialog box matches the following graphic, then click Save.

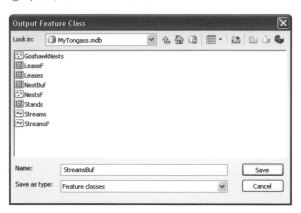

20 For the Buffer distance, click the Field option. Click the Field drop-down arrow and click Distance.

Distance is the field in the StreamsF layer table that contains the values 50 and 100. This time you won't dissolve the stream buffers as you did with the nest buffers.

21 Make sure your dialog box matches the following graphic, then click OK.

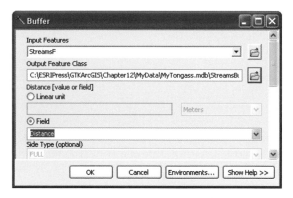

If necessary, click the Buffer link in the pop-up notification to see the tool's results report.

When the operation is completed, the new StreamBuf layer is added to the map. At the current scale, however, you can't get a good look at the stream buffers.

22 Close the ArcToolbox window. Click the Bookmarks menu and click Streams Closeup.

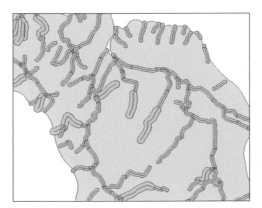

Now you can see the difference between the 50-meter buffers and the 100-meter buffers.

23 In the table of contents, right-click the StreamBuf layer and click Open Attribute Table.

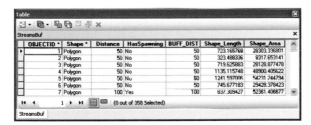

In addition to its standard attributes, the output table has attributes from the input table. (Because you did not dissolve the buffers, the output and input tables have the same number of records—there is one buffer for each stream. This correspondence makes it possible to copy attributes from one table to the other.)

24 Close the attribute table. Turn off the StreamsF layer. Turn on the NestBuf layer. In the table of contents, only the two buffer layers and the LeaseF layer should be turned on.

25 In the table of contents, right-click the LeaseF layer and click Zoom to Layer.

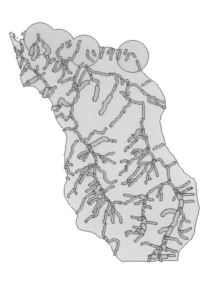

The two buffer layers define the areas within which trees cannot be cut. In the next exercise, you'll use overlays to define the areas in which they can be cut.

26 If you want to save your work, click the File menu and click Save As. Navigate to **\GTKArcGIS\Chapter12\MyData**. Rename the file **my_ex12a.mxd** and click Save.

27 If you are continuing with the next exercise, leave ArcMap open. Otherwise, exit the application. Click No if prompted to save your changes.

Overlaying data

A union overlay combines the features in two input layers to create a new dataset. In the following example, one polygon layer represents a land parcel and the other an oil spill. Part of the parcel lies outside the spill and part of the spill lies outside the parcel; in the middle, the two layers overlap. When the layers are unioned, the two original polygons become three. The area of overlap becomes a new feature, while the non-overlapping areas become multipart polygons.

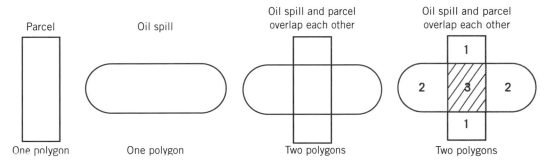

The oil spill and parcel layers are unioned to create three new polygons.

What kind of attributes does the union layer have? In addition to its own standard attributes, it includes all the attributes of both input layers. This doesn't mean, however, that every record has a value for each attribute. In this example, Feature 1 in the output table has no Spill_Type value because it is outside the oil spill, while Feature 2 has no Owner or Landuse values because it is outside the parcel. Only Feature 3, which spatially coincides with both input layers, has a value for every attribute.

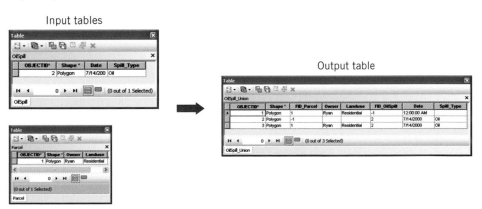

The output table contains the attributes of both input layers. Output features get the attributes values of input features with which they are spatially coincident.

10
11
12
13

A complication arises with the identifier attribute. One of the standard attributes of the union layer, assuming it is a geodatabase feature class, is the OBJECTID field, which assigns a unique ID number to each feature. (With other data formats, the identifier attribute is called FID or OID.) In this case, both of the input layers also have OBJECTID fields. ArcMap doesn't want to delete these fields—they may be useful to you—but at the same time, it doesn't want a table with three OBJECTID fields.

To resolve the conflict, the OBJECTID fields from the input layers are renamed in the union layer. The new name consists of the prefix "FID_" followed by the input layer name. Thus, the parcel layer's identifier attribute is renamed "FID_Parcel" and the oil spill layer's is renamed "FID_OilSpill."

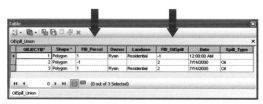

Renamed OBJECTID fields

In an intersect overlay, the process is a little simpler, since only the area of overlap is preserved. An Intersect of the oil spill and parcel layers would consist of a single feature:

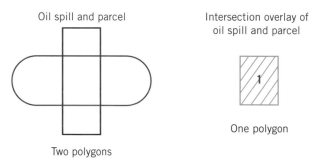

As with a union layer, an intersect layer includes its own standard attributes plus all the attributes of both input layers. In an intersect, each attribute is populated for every record.

A union overlay requires that both input layers be polygon layers. In an intersect, the input layers can either be two polygon layers or a polygon layer and a line layer. In the latter case, the output is a line layer.

Exercise 12b

In this exercise, you'll union the nest and stream buffer layers from the previous exercise to create a single layer of the land that cannot be harvested. Then you'll union this layer with a layer of stands in lease F. Because new features will be created wherever stand polygons and buffer polygons overlap, every output feature will lie either entirely inside or entirely outside a buffer. The set of features lying outside buffers represents harvestable land.

1 In ArcMap, open **ex12b.mxd** from the **C:\ESRIPress\GTKArcGIS\Chapter12** folder.

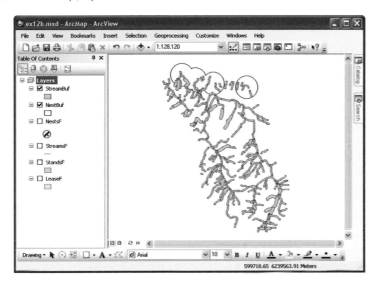

The map shows the buffers for streams and goshawk nests. The other layers are turned off.

2 On the Standard toolbar, click the ArcToolbox window button.

3 In the ArcToolbox window, expand the Analysis Tools if necessary, then click the plus sign next to Overlay.

4 Double-click the Union tool.

You'll specify the layers to overlay and designate an output feature class.

5 Click the Input Features drop-down arrow and click NestBuf.

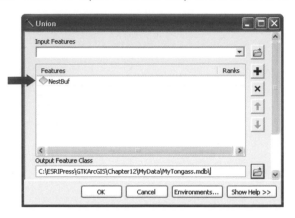

The selected layer is added to the list of layers that will be processed in a union.

6 Click the Input Features drop-down arrow again and click StreamBuf.

7 Click the Browse button next to the Output Feature Class box. In the Output Feature Class dialog box, navigate to **\GTKArcGIS\Chapter12\MyData** and double-click **MyTongass.mdb**.

8 In the Name box, type **NoCutArea**. Make sure your dialog box matches the following graphic, then click Save.

By default, the output layer attribute table contains its standard attributes plus the attributes from both input layers. Optionally, you can omit the input layer identifier attributes. Alternatively, you can omit all input attributes except the identifiers. This creates a smaller output table and is convenient if you don't need to work with attributes. (If you later decide that you do need the input attributes, you can join or relate back to the input layer tables using the common identifier attribute.)

Right now, all you need are the identifier attributes.

9 Click the JoinAttributes drop-down arrow and click ONLY_FID. (You might have to scroll down in the dialog box to see it.)

10 Make sure your dialog box matches the following graphic, then click OK.

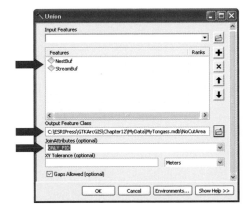

When the operation is completed, the NoCutArea layer is added to the map. The new layer, consisting of all buffered areas from the two input layers, defines the zone where no trees may be harvested.

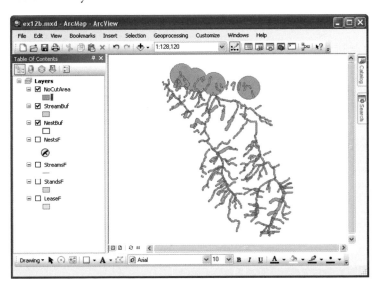

11 In the table of contents, right-click the NoCutArea layer and click Open Attribute Table.

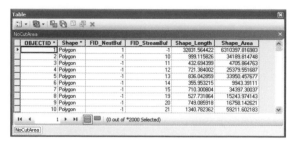

The table contains the four standard polygon feature class attributes (OBJECTID, Shape, Shape_Length, and Shape_Area). It also contains the renamed identifier attributes from the input layers: FID_NestBuf and FID_StreamBuf.

In the exercise introduction, you saw that attributes are not populated for every record in a Union attribute table. Identifier attributes, however, are completely populated. Every record in the Attributes of NoCutArea has not only an OBJECTID value, but also an FID_StreamBuf value and an FID_NestBuf value—regardless of which input feature is spatially coincident with the output feature.

For example, the second record in the NoCutArea table has an OBJECTID value of 2. This value is the feature's new identifier. The record's FID_StreamBuf value is 10. This means that the feature coincides spatially with the feature that has the OBJECTID of 10 in the StreamBuf table. The record's FID_NestBuf value is -1. The -1 value means that the feature does not coincide spatially with any feature in the NestBuf layer. Notice that the first record has an FID_StreamBuf value of -1. This means that the feature does not coincide spatially with any feature in the StreamBuf layer.

Later in this exercise, you will see how to use this information to your advantage.

12 Close the table. Turn on the StandsF layer.

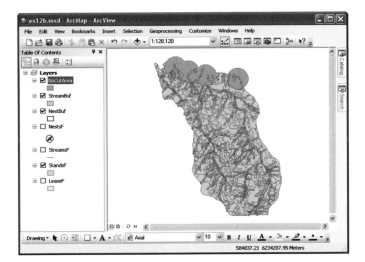

To find the harvestable land, you will union the NoCutArea layer with StandsF. After the overlay, you will be able to select the polygons that represent harvestable areas.

13 In the ArcToolbox window, double-click the Union tool.

14 Click the Input Features drop-down arrow and click NoCutArea to add it to the list of input layers. Do the same for StandsF.

15 Click the Browse button next to the Output Feature Class box. In the Output Feature Class dialog box, navigate to **\GTKArcGIS\Chapter12\MyData** and double-click **MyTongass.mdb**.

16 In the Name box, type **Final**. Make sure your dialog box matches the following graphic, then click Save.

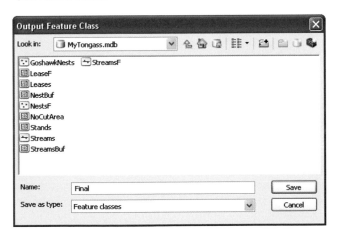

Leave the Join Attributes drop-down list set to ALL. This time you will include all
input attributes in the Union layer. (You will need attributes like StandValue and
ValuePerMeter to recalculate the stand values in the next exercise.)

17 Make sure your dialog box matches the following graphic, then click OK.

When the operation is completed, the Final layer, which has more than 5,000 features,
is added to the map.

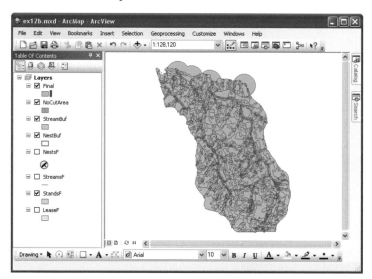

At this scale, it's hard to tell if you're looking at the result of a spatial analysis or just
a plate of spaghetti. In a moment, you'll zoom in, but first you'll look at the attribute
table of the Final layer.

Exercise **12b** Overlaying data

18 In the table of contents, right-click the Final layer and click Open Attribute Table.

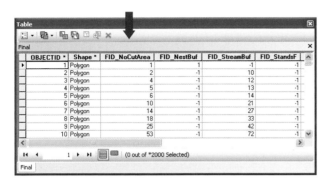

OBJECTID	Shape	FID_NoCutArea	FID_NestBuf	FID_StreamBuf	FID_StandsF
1	Polygon	1	1	-1	-1
2	Polygon	2	-1	10	-1
3	Polygon	4	-1	12	-1
4	Polygon	5	-1	13	-1
5	Polygon	6	-1	14	-1
6	Polygon	10	-1	21	-1
7	Polygon	14	-1	27	-1
8	Polygon	18	-1	33	-1
9	Polygon	25	-1	42	-1
10	Polygon	53	-1	72	-1

The FID_NoCutArea attribute is the renamed OBJECTID from the NoCutArea layer. If a record has a value other than -1 in this field, it means the output feature coincides spatially with a buffer feature; in other words, it's not harvestable. If a record has the value -1, it means the output feature does not coincide with a buffer and therefore is harvestable.

The harvestable area of lease F, therefore, is the area composed of all polygons in the Final layer that have the value -1 in the FID_NoCutArea field.

19 Close the Attributes of Final table. Close ArcToolbox. In the table of contents, turn off all layers except Final.

20 Click the Bookmarks menu and click Close-up.

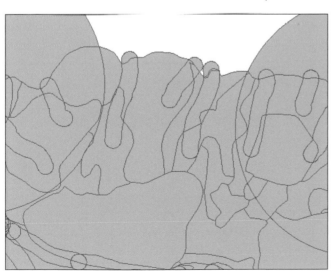

To see which areas can be harvested, you will turn on labels.

21 In the table of contents, double-click the Final layer. On the Layer Properties dialog box, click the Labels tab. Check the "Label features in this layer" check box. Click the Label Field drop-down arrow and click FID_NoCutArea. Make sure your dialog box matches the following graphic, then click OK.

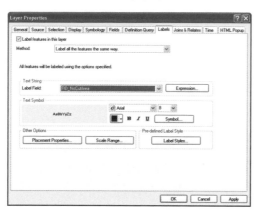

On the map, polygons labeled -1 represent harvestable areas.

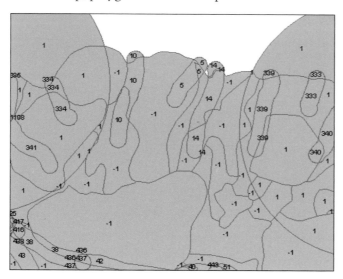

It would be easier to read the map if you used colors instead of labels. Applying symbology is not part of this exercise, but you are welcome to do it on your own, using what you learned in chapter 5. Your result might look like this:

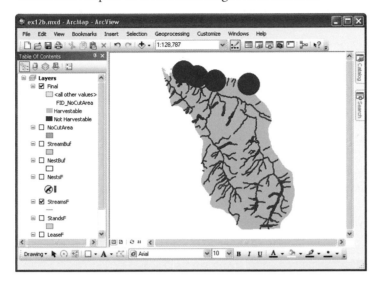

In the next exercise, you will recalculate the Final layer's StandValue attribute to get an accurate total value for harvestable land.

22 If you want to save your work, save it as **my_ex12b.mxd** in the **\GTKArcGIS\Chapter12 \MyData** folder.

23 If you are continuing with the next exercise, leave ArcMap open. Otherwise, exit the application. Click No if prompted to save your changes.

Calculating attribute values

You can write an expression to calculate attribute values for all records in a table or just for selected ones. For numeric attributes, the expression can include constants, functions, or values from other fields in the table. For text attributes, the expression can include character strings that you type or text values from other fields.

Exercise 12c

The graph you made in chapter 11 showed the timber value of lease F to be about 1.5 billion dollars. In this exercise, you'll adjust that value to take into account only harvestable areas. You'll create a definition query to display these areas, then you'll recalculate stand values to determine how much the total harvestable area is worth.

1   In ArcMap, open **ex12c.mxd** from the **C:\ESRIPress\GTKArcGIS\Chapter12** folder.

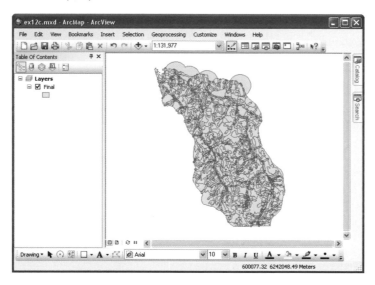

The map contains the Final layer you created in the last exercise.

In the previous exercise, you saw that harvestable areas have the value -1 in the FID_NoCutArea field. Using that value, you will create a definition query to display only the harvestable areas.

2   In the table of contents, double-click the Final layer. In the Layer Properties dialog box, click the Definition Query tab.

3 Click Query Builder to open the Query Builder dialog box.

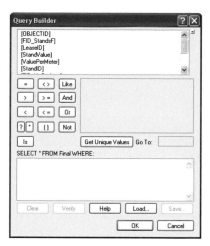

4 In the Fields box, scroll down and double-click [FID_NoCutArea] to add it to the expression box. Click the equals (=) button, then click Get Unique Values. In the Unique Values list, double-click -1. Make sure your expression matches the following graphic, then click OK.

The query displays in the Definition Query box.

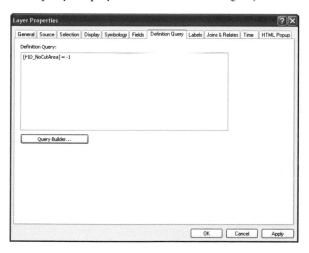

5 Click OK on the Layer Properties dialog box.

On the map, only features satisfying the query are displayed. The layer attribute table will show only the records corresponding to these features.

Now you'll update the StandValue attribute for these features.

6 In the table of contents, right-click the Final layer and click Open Attribute Table.

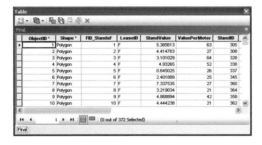

The table shows 372 records, rather than the original 5,439.

The values in the StandValue field need to be updated. Although some of the original stand polygons were preserved intact in the Final layer (and still have correct stand values), many other stands—all those that were overlapped by a nest or stream buffer—were split in the last overlay. The resulting smaller polygons have correct area values because ArcMap automatically updates the Shape_Area attribute, but their stand values, which were simply copied over from the StandsF table, are wrong. To correct the stand values, you will multiply the area of each feature by its value per meter.

7 Right-click the field name StandValue and click Field Calculator. If necessary, click Yes on the message warning you that you cannot undo the calculation.

The Field Calculator dialog box opens.

8 In the Fields box, double-click Shape_Area to add it to the expression box. Click the multiplication (*) button. Again in the Fields box, double-click ValuePerMeter.

This expression will give you the updated stand values in dollars. In the table, however, the stand values are expressed in millions of dollars.

9 Click at the beginning of the expression and type an opening parenthesis "(" followed by a space. Click at the end of the expression and type a space followed by a closing parenthesis ")". Click the division (/) button. Type a space and type **1000000**. Make sure your expression matches the one in the following graphic, then click OK.

The values in the StandValue field are recalculated.

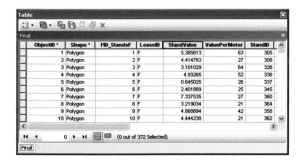

10 Right-click the StandValue field name and click Statistics.

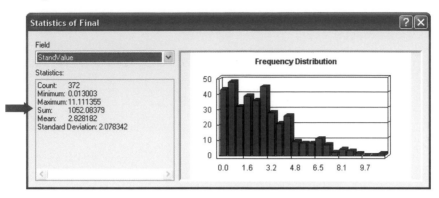

The sum of the values is 1,052.08379. The harvestable value of lease F is therefore just over a billion dollars—about two thirds of the original calculation shown in your graph from exercise 11b, step 9.

11 Close the Statistics window and the table.

Your company will use this information to make a competitive bid. It's a big investment, but tree harvesting is expensive. You have to move heavy equipment into the area, supply the labor force, and construct roads.

A more detailed analysis would consider additional factors, such as the locations of existing roads, the slope of the land, and other protected areas like stands of old-growth trees.

12 If you want to save your work, save it as **my_ex12c.mxd** in the **\GTKArcGIS\Chapter12 \MyData** folder.

In the next chapter, you'll launch ArcMap from ArcCatalog. So even if you are continuing, you should exit ArcMap now.

13 Close ArcMap. Click No if prompted to save your changes.

Building a spatial model

In chapters 11 and 12, you carried out many spatial analysis operations, such as Dissolve, Clip, Buffer, and Union, using ArcToolbox. These operations were not ends in themselves, but steps in a larger analytical process. Before undertaking a similarly complicated GIS project, you might find it useful to draw a flowchart or diagram that identifies the goal and the analytical steps that lead to it. What data will you need? What geoprocessing will be required? Which outputs will become inputs to new operations?

In ArcGIS Desktop 10, ModelBuilder is available to help you do that. ModelBuilder provides a design window where spatial analysis operations can be defined, sequentially connected, and carried out, all with the use of drag-and-drop icons. ModelBuilder is both a workflow diagramming tool and a processing environment. It keeps track of the operations that you run, their results, and their interdependencies. It gives you a convenient way to build a large project from its component parts and run its processes separately or together. Your model can be changed at any time to incorporate new data, add new conditions, or try different assumptions (best case, worst case, what if…).

Chapter 20 in this book revisits the Tongass Forest Lease analysis and shows you how to design and execute the same project in ModelBuilder.

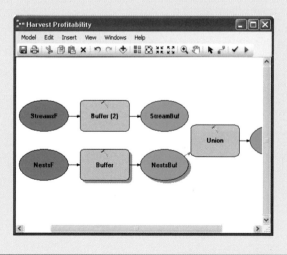

Chapter 13

Projecting data in ArcMap

Projecting data on the fly

Defining a projection

Locations on the earth's surface are defined with reference to lines of latitude and longitude. Latitude lines, or parallels, run parallel to the equator and measure how far north or south you are of the equator. Longitude lines, or meridians, run from pole to pole and measure how far east or west you are of the prime meridian (the meridian that passes through Greenwich, England).

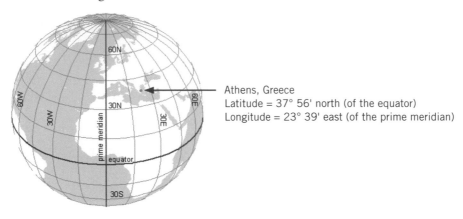

Athens, Greece
Latitude = 37° 56' north (of the equator)
Longitude = 23° 39' east (of the prime meridian)

The mesh of intersecting parallels
and meridians is called a graticule.

Latitude and longitude are measurements of angles, not of distances. Latitude is the angle between the point you are locating, the center of the earth, and the equator. Longitude is the angle between the prime meridian, the center of the earth, and the meridian on which the point you are locating lies. Because latitude and longitude are angles, their values are expressed in degrees, minutes, and seconds.

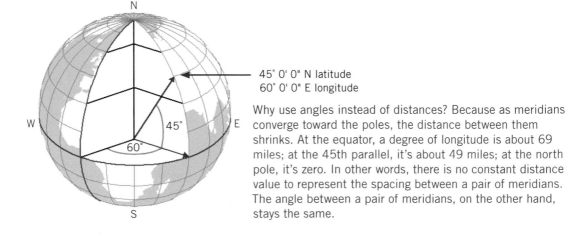

45° 0' 0" N latitude
60° 0' 0" E longitude

Why use angles instead of distances? Because as meridians converge toward the poles, the distance between them shrinks. At the equator, a degree of longitude is about 69 miles; at the 45th parallel, it's about 49 miles; at the north pole, it's zero. In other words, there is no constant distance value to represent the spacing between a pair of meridians. The angle between a pair of meridians, on the other hand, stays the same.

Starting from the equator, latitude values go to +90° at the north pole and to −90° at the south pole. Starting from the prime meridian, longitude values go to +180° eastward and to −180° westward. (The +180° and −180° meridians are the same.)

Latitude and longitude are the basis of a geographic coordinate system, a system that defines locations on the curved surface of the earth. Because different estimates have been made of the earth's shape and size, there are a number of different geographic coordinate systems in use. Although they are similar, the precise latitude–longitude coordinates assigned to locations vary from one system to the next.

Some geographic coordinate systems are based on the assumption that the earth is a sphere. This simple model is adequate for many purposes.

Most geographic coordinate systems are based on the assumption that the earth is a spheroid. (A spheroid is to a sphere as an oval is to a circle.) This is a more accurate model because the earth bulges slightly at the equator and is flattened at the poles. Historically, several spheroids of varying dimensions have been calculated. (This picture greatly exaggerates the flattening, which is really only a fraction of a percent.)

To make a map, the earth must be represented on a flat surface. This is accomplished by a mathematical operation called a map projection.

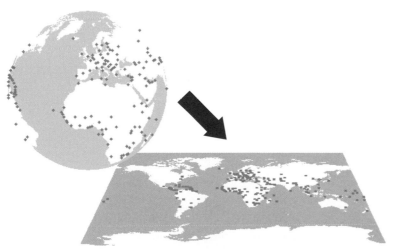

A map projection flattens the earth. Locations on the earth are systematically assigned to new positions on the map. This can be done by many different methods (infinitely many, in fact).

Locations on a map are defined with reference to a grid of intersecting straight lines. One set of lines runs parallel to a horizontal x-axis; the other set runs parallel to a vertical y-axis. The coordinates of any point are expressed as a distance value along the x-axis and a distance value along the y-axis (from the intersection of the axes).

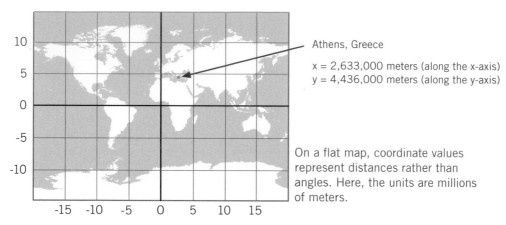

Athens, Greece

x = 2,633,000 meters (along the x-axis)
y = 4,436,000 meters (along the y-axis)

On a flat map, coordinate values represent distances rather than angles. Here, the units are millions of meters.

A projected coordinate system, a system that defines locations on a flat map, is based on x,y coordinates. The x,y coordinates of any given point, such as Athens, depend on the map projection being used, the units of measure (meters, feet, or something else), and on where the map is centered. If Athens is made the center of the map, for example, its x,y coordinates will be 0,0. Thus, the number of possible projected coordinate systems is unlimited.

Since the world is more or less round and maps are flat, you can't go from one to the other without changing the properties of features on the surface. Every map projection distorts the spatial properties of shape, area, distance, or direction in some combination.

The Mercator projection preserves shape but distorts area. On the map, Greenland is much larger than Brazil, but on the earth it is smaller.

The sinusoidal projection preserves area but distorts shape. The proportional sizes of Greenland and Brazil are correct, but not their shapes.

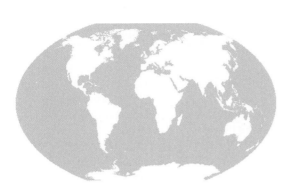

The Winkel tripel projection balances distortion. No single property is faithfully preserved, but none is excessively distorted.

The azimuthal equidistant projection preserves true distance and direction from a single point (in this case, Athens) to all other locations on the map.

Your choice of map projection lets you control the type of distortion in a map for your area of interest. If you are mapping an area the size of a small country, and using an appropriate projection, the effects of distortion will be insignificant. If you are mapping the whole world, distortion will always be noticeable, but you can reduce or eliminate certain types of distortion according to your purpose.

For more information about coordinate systems and map projections, click the Contents tab in ArcGIS Desktop Help and navigate to *Professional Library > Guide books > Map projections > Projected coordinate systems > About map projections.*

10
11
12
13

Projecting data on the fly

Every spatial dataset has a coordinate system. If it's a geographic coordinate system, its features store latitude–longitude values. If it's a projected coordinate system, its features store x,y values.

Besides storing feature coordinates, a dataset contains other coordinate system information. The definition of a geographic coordinate system, for example, includes the dimensions of the sphere or spheroid it's based on and other details. The definition of a projected coordinate system includes the projection it's based on, the measurement units, and other details. Each projected coordinate system is also associated with a particular geographic coordinate system, since its x,y coordinates were at some time projected from a set of latitude–longitude coordinates.

You can find the coordinate system of a dataset in the Description tab in ArcCatalog (to view its metadata). You can also find it in the XY Coordinate System tab of the Properties dialog box (right-click a feature and click Properties.) Alternatively, find it in ArcMap by clicking the Source tab of the Layer Properties dialog box.

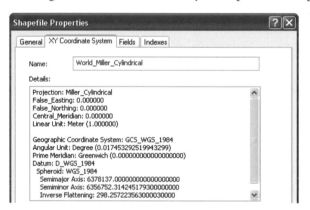

This World Countries shapefile has a projected coordinate system. You also see the geographic coordinate system from which the projected coordinates were derived (WGS 1984).

When you first add a layer to a data frame in ArcMap, it displays according to the coordinate values of its features—geographic or projected, as the case may be.

What happens when you add a second layer to the data frame? If it has the same coordinate system as the first layer, there's no problem. But what if its coordinate system is different? Since coordinates tell ArcMap where to draw features, there is a potential conflict.

ArcMap resolves this conflict automatically. You already know that a map projection is a math operation that changes geographic coordinates to projected coordinates. ArcMap knows the equations (forward and backward) for hundreds of projections. So when the second layer you add has a different coordinate system from the first, ArcMap changes Layer 2's coordinates to match Layer 1's. This process is called on-the-fly projection.

How does it work? Suppose you add a layer of world countries to a data frame. Say it's in the Miller cylindrical projected coordinate system. ArcMap stores the information about this coordinate system (and the geographic coordinate system it's based on) as a property of the data frame. Suppose you next add a layer of world capitals in the sinusoidal projected coordinate system. ArcMap checks the data frame properties and knows it can't display this layer according to its sinusoidal coordinates—it has to change them to Miller cylindrical coordinates. To do this, it looks at the geographic coordinate system that the world capitals layer is based on, "unprojects" the sinusoidal coordinates to latitude–longitude coordinates, and then projects these latitude–longitude coordinates to Miller cylindrical coordinates. The result is that the capitals display in their correct relationship to the countries.

Now suppose you add a third layer of world rivers that is in a geographic coordinate system. The process is simpler. All ArcMap has to do is project the latitude–longitude coordinates to Miller cylindrical coordinates. All three layers now occupy the same "coordinate space" and display in correct spatial alignment, even though the coordinates stored with the features are different for each layer.

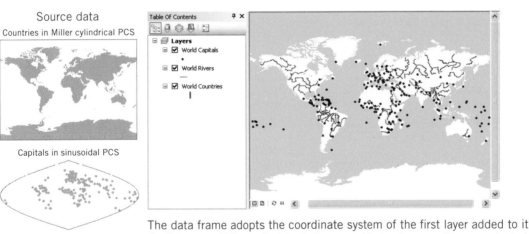

Source data

Countries in Miller cylindrical PCS

Capitals in sinusoidal PCS

Rivers in GCS

The data frame adopts the coordinate system of the first layer added to it (in this case, Miller cylindrical). Layers added subsequently are projected on the fly to this coordinate system.

You don't have to use the coordinate system of the first layer you add— this is just the ArcMap default. You can assign any coordinate system you want to the data frame (even one that none of the layers have) and all layers will be projected to match it.

On-the-fly projection doesn't change the feature coordinates of the dataset on disk, or the dataset's coordinate system information. On-the-fly projection has effect only within a single data frame. If you were to add these same three layers to another data frame in the same map document—but if you added the world capitals layer first—all the layers would display in the sinusoidal projected coordinate system.

On-the-fly projection works best when layers are based on the same geographic coordinate system. If Layer 2 has a different geographic coordinate system from Layer 1, you'll get a message like this:

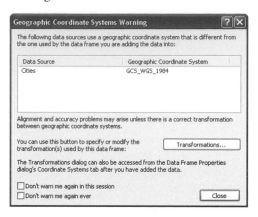

The message tells you that ArcMap can display the layer, but that the spatial alignment probably won't be just right. That's because ArcMap can't convert one geographic coordinate system to another without some help from you. Without that help, the data will be misaligned to the extent that the geographic coordinate systems differ from each other. This difference is usually too small to notice at the scale of a world map, or even a continental map. But if you are making a map that demands highly accurate feature positioning, or if you are geoprocessing the data, you will first want to convert one layer's geographic coordinate system to that of the other.

This operation is called a geographic coordinate system transformation (or a datum transformation). ArcMap supports it, but the process is not described in this book. For more information, click the Contents tab in ArcGIS Desktop Help and navigate to *Professional Library > Guide books > Map projections > Geographic transformations > Geographic transformation methods.*

Exercise 13a

You work for the U.S. Census Bureau and are creating a map of the United States that shows population change between 1990 and 2000. The map will have three data frames: one for the lower forty-eight states, one for Alaska, and one for Hawaii. The same shapefile of U.S. states (in a geographic coordinate system) will be used in all three data frames. You'll set a different projected coordinate system for each data frame and ArcMap will project the data on the fly.

Why use three data frames instead of just one? Showing all fifty states in the same data frame would mean including all the space that separates Hawaii and Alaska from the contiguous forty-eight states. You would have to zoom out so far that little states would be hard to see. By putting Alaska and Hawaii in their own data frames, however, you can zoom in on them and move them close to the other forty-eight states.

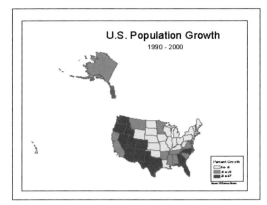 

The United States drawn in a single data frame.

The United States drawn in three data frames. Alaska and Hawaii aren't shown in true proportion or at true distance from the lower forty-eight states, but that isn't important to the purpose of your map.

You will use three versions of the Albers equal area conic projection. This is an excellent general-purpose projection for areas in middle latitudes, especially those having an east—west orientation. It preserves the spatial property of area, which means that map features are displayed at their true proportional size. Shapes are only minimally distorted for an area the size of the United States.

1 Start ArcCatalog. In the ArcCatalog tree, double-click the connection to **C:\ESRIPress \GTKArcGIS**. Double-click the **Chapter13** folder. Double-click the **Data** folder.

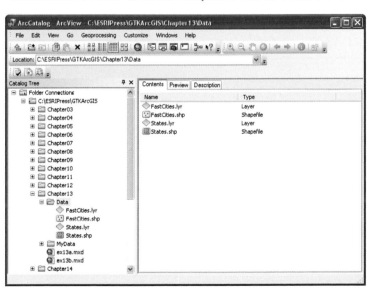

There are two shapefiles: FastCities and States. Each has a corresponding layer file. In this exercise, you'll use the States data. (You'll work with FastCities, which contains the fastest-growing city in each state, in the next exercise.)

2   In the catalog window, click **States.lyr** and click the Preview tab.

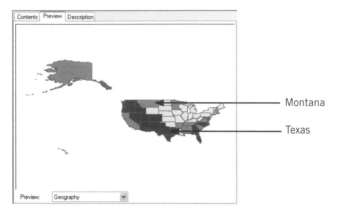

The layer has been symbolized with a color ramp that shows population change. The states, displayed according to their geographic coordinates, are distorted both in shape and area. The northern boundary of the United States appears to be a straight line, for instance, while Alaska looks stretched out. Montana seems to be roughly the same size as Texas, while in reality Texas is nearly twice as large.

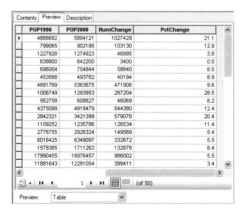

3   Click the Preview drop-down arrow and click Table. Scroll to the right.

The table has 1990 and 2000 population figures for each state, as well as the population change in raw numbers and percentages. Every state grew during the decade. The layer is classified and symbolized on the PctChange attribute.

You'll confirm that the States shapefile has a geographic coordinate system.

4 In the catalog tree, right-click **States.shp** and click Properties.

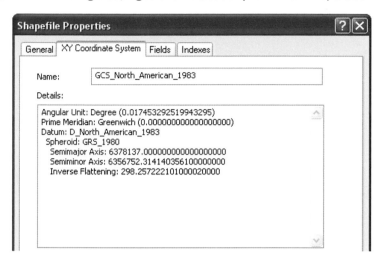

The Shapefile Properties window shows a geographic coordinate system called GCS_ North_American_1983. (This is a common geographic coordinate system for spatial data covering the United States. It is also called NAD83 or North American Datum of 1983.)

Next, you'll start ArcMap and add the States layer file to a map document.

5 Click Cancel to close the properties window. In the catalog tree, double-click **ex13a.mxd**.

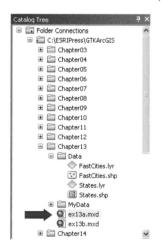

ArcMap opens in layout view.

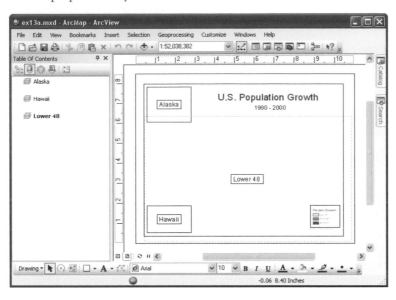

The map contains three empty data frames: Alaska, Hawaii, and Lower 48. Lower 48 is active. A title and legend have already been added to the map.

You'll add the States layer to each data frame, zoom in appropriately, and apply map projections to the data frames.

6   Position the ArcCatalog and ArcMap windows so you can see the catalog tree and the ArcMap table of contents.

7   In the catalog tree, click **States.lyr** and drag it to the bottom of the ArcMap table of contents. Click the ArcMap title bar to bring ArcMap forward.

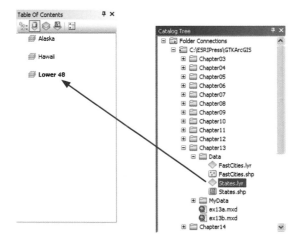

The layer is added to the table of contents and displays in the data frame. It looks as it did when you previewed it in ArcCatalog. You'll use a bookmark to zoom in to the lower forty-eight states.

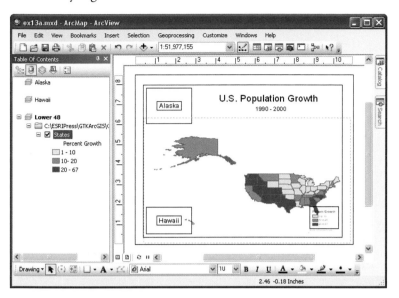

8   Click the Bookmarks menu and click Lower 48.

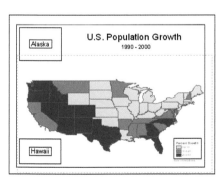

10
11
12
13

9 In the table of contents, double-click the Lower 48 data frame. In the Data Frame Properties dialog box, click the Coordinate System tab.

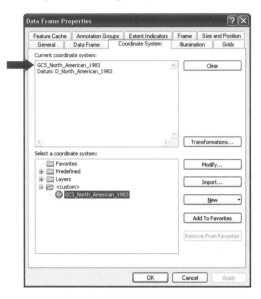

By default, the data frame is set to the coordinate system of the first layer added to it. As you saw in step 5, that is GCS_North_American_1983.

10 In the Select a coordinate system box, click the plus sign next to Predefined.

11 Click the plus sign next to Projected Coordinate Systems, the plus sign next to Continental, then the plus sign next to North America.

A variety of projected coordinate systems for North American data is listed.

12 Click USA Contiguous Albers Equal Area Conic. (You can hold the mouse pointer over a name to see all of it.) The details of the projection appear in the Current coordinate system box. Make sure that your dialog box matches the following graphic, then click OK.

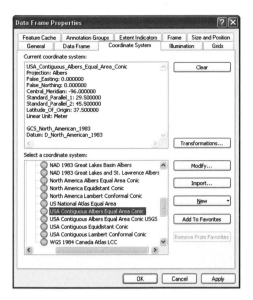

The layer is projected on the fly. You'll use a bookmark to zoom in closer.

13 Click the Bookmarks menu and click Lower 48 after Projection.

The states are displayed in the Albers equal area projection.

Now you'll copy the States layer to the other two data frames and set their coordinate system properties.

14 In the table of contents, right-click the States layer and click Copy. In the table of contents, right-click the Hawaii data frame and click Paste Layer(s). Again in the table of contents, right-click the Alaska data frame and click Paste Layer(s).

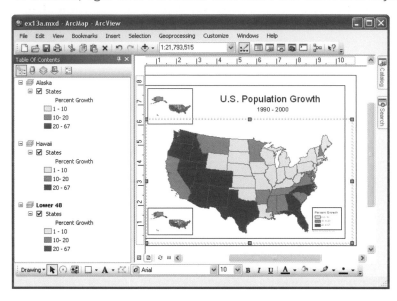

The States layer now displays in all three data frames.

15 In the table of contents, right-click the Alaska data frame and click Activate.

16 In the table of contents, double-click the Alaska data frame. In the Data Frame Properties dialog box, click the Coordinate System tab if necessary.

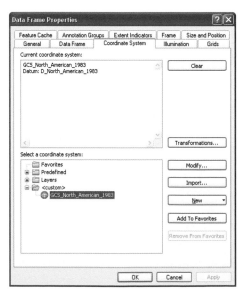

As before, the data frame's coordinate system is set to GCS_North_American_1983. You'll set an Albers projection developed for Alaska.

17 In the Select a coordinate system box, click the plus sign next to Predefined, the plus sign next to Projected Coordinate Systems, the plus sign next to Continental, and the plus sign next to North America.

18 Click Alaska Albers Equal Area Conic. Make sure that your dialog box matches the following graphic, then click OK.

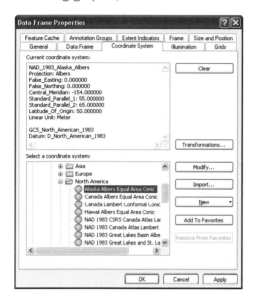

The layer is projected on the fly. Notice that the lower forty-eight states are set at an odd angle. The projection minimizes distortion for Alaska, but other areas may be severely distorted. You won't show these areas in the data frame, so it doesn't matter.

19 Click the Bookmarks menu and click Alaska.

20 In the table of contents, right-click the Hawaii data frame and click Activate.

21 Following the same steps you used for the other data frames, set the coordinate system of the Hawaii data frame to Hawaii Albers Equal Area Conic.

22 Click the Bookmarks menu and click Hawaii.

Each of the three data frames now has a different projected coordinate system. (All use the Albers equal area conic projection, but the projection settings are customized in each case to represent the area of interest as accurately as possible.)

23 If you want to save your work, click the File menu and click Save As. Navigate to **\GTKArcGIS\Chapter13\MyData**. Rename the file **my_ex13a.mxd** and click Save.

24 If you are continuing with the next exercise, keep ArcMap open; otherwise, exit the application. Click No if prompted to save your changes.

25 Close ArcCatalog.

Defining a projection

Features in a dataset never lose their coordinates, but the dataset may not have the information that identifies its coordinate system. This can happen in particular with shapefiles, where the coordinate system information is stored as a separate file (with the extension .prj).

If you add a layer that is missing its coordinate system information, ArcMap will look at the feature coordinate values before proceeding. If it sees that all the values are between 0 and 180, it knows that the data is in a geographic coordinate system. It won't know which one, but it will make a default assumption that enables it to display the layer with other layers already in the data frame. (The spatial alignment probably won't be exact—just as when you add layers that have different geographic coordinate systems—but it will usually be acceptable for small-scale maps.)

If ArcMap sees that the coordinate values are big six- or seven-digit numbers, it knows that the data is in a projected coordinate system, but again, it doesn't know which one. You'll get a message similar to this one:

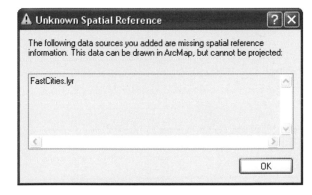

The message tells you that the layer can't be projected on the fly. In other words, ArcMap can add the layer to the data frame, but it can't change the layer's coordinates to match the data frame's coordinate system. The result is often a serious display problem, because very different sets of coordinates are trying to fit in the same map space. For example, if the data frame is set to a geographic coordinate system, the unknown layer probably won't be visible (its projected coordinates are too far apart from latitude–longitude values). If the data frame is set to a projected coordinate system, the unknown layer may display, but will probably be seriously misaligned. Sometimes you get lucky, though. If the data frame happens to have the same projected coordinate system as the unknown layer, the data will line up correctly on its own.

If a dataset is missing its coordinate system information, you should try to find out what it is. This may involve contacting the data provider or looking through files for supporting documentation. Once you know the coordinate system, you can assign it to the dataset using ArcToolbox. This process is called defining a projection.

Exercise 13b

In this exercise, you'll add a layer of the fastest-growing cities in each state to your map. From a colleague at the Census Bureau, you have acquired a shapefile of projected data along with a corresponding layer file. Unfortunately, the shapefile is missing its projection (.prj) file. This means that ArcMap can't recognize the coordinate system and is unable to project the data on the fly.

After e-mailing your colleague, you have received the following reply:

"Sorry that the projection wasn't defined. The FastCities shapefile is in the North America Lambert conformal conic projected coordinate system."

You'll add this coordinate system information to the shapefile with the Define Projection tool in ArcToolbox.

1   In ArcMap, open **ex13b.mxd** from **C:\ESRIPress\GTKArcGIS\Chapter13**.

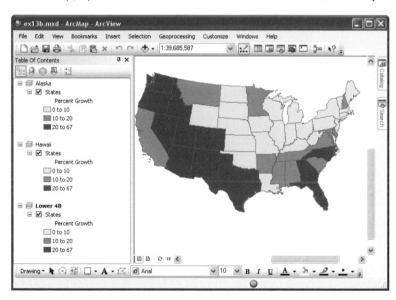

The map opens in data view. The Lower 48 data frame is active.

2   On the Standard toolbar, click the Add Data button.

3 In the Add Data dialog box, navigate to **\GTKArcGIS\Chapter13\Data** and click **FastCities.lyr**, as shown in the following graphic. Click Add.

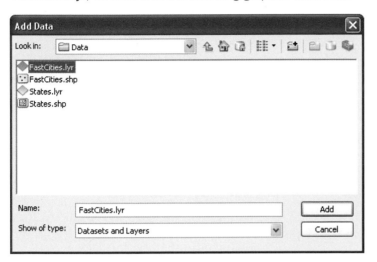

ArcMap warns you that the layer is missing spatial reference information.

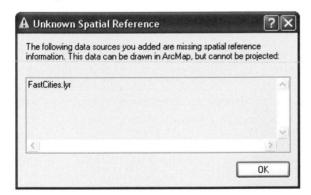

If you don't see the warning, that's okay. In your ArcGIS past, you may have come across the Warning dialog box shown on page 346 and checked its Don't warn me again ever box. If you would like coordinate system warnings to appear again, run the AdvancedArcMapSettings.exe executable file, which is located in the Utilities folder of your ArcGIS installation. On the Advanced ArcMap Settings dialog box, click the Miscellaneous tab and uncheck Skip datum check.

4 Click OK on the warning message.

ArcMap adds the FastCities layer. At first glance, things appear normal. If you take a closer look, though, you'll see that some cities are lying outside the U.S. boundary.

The data frame's current coordinate system is the USA contiguous Albers equal area conic (the one you set in the previous exercise). From your colleague, you now know that the coordinate system of the FastCities layer is North America Lambert conformal conic. ArcMap, however, does not know this.

Because it couldn't project the FastCities layer on the fly, ArcMap simply drew the features where their Lambert coordinates said they should go. But since these Lambert numbers were being plugged into Albers coordinate space, the locations are wrong. (It's sort of like digging for buried treasure in the right place on the wrong island.)

As it happens, the Albers and Lambert coordinates for the United States are not hugely different; therefore, although the layers don't align correctly, you can at least see them both together.

You'll zoom in for a closer look.

5 Click the Data View button. Then click the Bookmarks menu and click Southeast.

Among other problems, Myrtle Beach is in the ocean; Grand Rapids, Michigan, is in Indiana; Columbus, Ohio, is in Kentucky; and Nashville, Tennessee, is in Alabama.

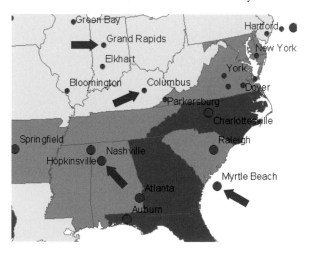

Fortunately, you can use ArcToolbox to add the missing coordinate system information. Once you do that, ArcMap will be able to reproject the FastCities layer, and you'll see the cities display in their correct locations.

6 On the Standard toolbar, click the ArcToolbox window button.

7 In the ArcToolbox window, click the plus sign next to Data Management Tools. Click the plus sign next to Projections and Transformations.

8 Double-click the Define Projection tool to open it.

9 Click the Input Dataset or Feature Class drop-down arrow and click FastCities.lyr.

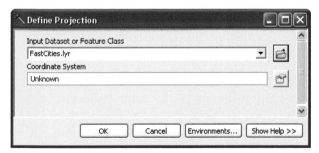

You see that the layer's coordinate system is Unknown.

10 Click the Properties button next to the Coordinate System box.

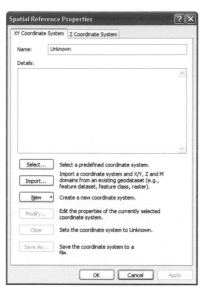

In the Spatial Reference Properties dialog box, you can select from coordinate systems already defined by ArcGIS, import a coordinate system from another dataset, or create an entirely new coordinate system. (You might do this, for example, if you were making a map of a newly discovered planet, since ArcGIS only has predefined coordinate systems for planets in our solar system.)

The Lambert conformal conic for North America, however, is defined by ArcGIS.

11 Click Select.

12 In the Browse for Coordinate System dialog box, double-click the Projected Coordinate Systems folder, the Continental folder, and the North America folder. Click North America Lambert Conformal Conic.prj. (You may need to widen the Name header to see the full projection names.) Make sure that your dialog box matches the following graphic, then click Add.

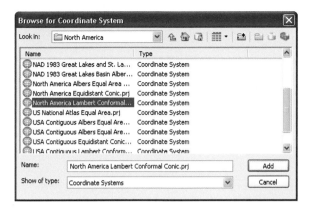

The Spatial Reference Properties dialog box is updated with the coordinate system details.

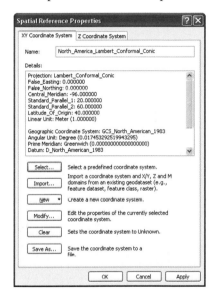

13 Click OK.

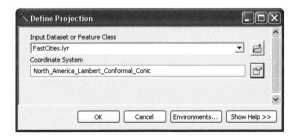

The coordinate system information will be written to a projection file (.prj) associated with the FastCities shapefile.

14 Click OK.

The Define Projection tool runs and you see a progress report.

15 Click Close on the progress report. Close the ArcToolbox window.

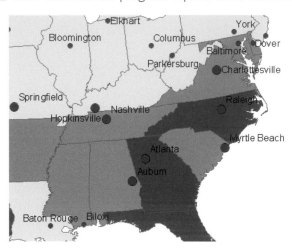

With the correct coordinate system information, ArcMap reprojects the FastCities layer into the Albers equal area conic projection. The cities appear in their correct positions.

16 From the Bookmarks menu, click Lower 48 after Projection.

You will copy the FastCities layer to the other two data frames.

17 On the View menu, click Layout View.

18 In the table of contents, right-click FastCities.lyr and click Copy.

19 In the table of contents, right-click the Hawaii data frame and click Paste Layer(s). Again in the table of contents, right-click the Alaska data frame and click Paste Layer(s).

20 Click the Layout View button.

Honolulu and Anchorage appear in their respective data frames.

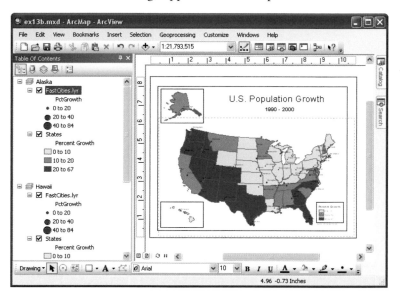

In the Alaska data frame, the FastCities layer is reprojected to Alaska Albers equal area conic. In the Hawaii data frame, it is reprojected to Hawaii Albers equal area conic.

Your map shows which states and cities have experienced the most growth between 1990 and 2000.

21 If you want to save your work, save it as **my_ex13b.mxd** in the **\GTKArcGIS\Chapter13 \MyData** folder.

In the next chapter, you will work exclusively with ArcCatalog, so you'll exit ArcMap now.

22 Close ArcMap. Click No if prompted to save your changes.

Chapter 14

Building geodatabases

Creating a personal geodatabase
Creating feature classes
Adding fields and domains

Spatial data comes in a variety of formats that can be managed in ArcCatalog and added as layers to ArcMap. Many of these formats, including shapefiles, coverages, CAD files, and geodatabases, organize spatial data into feature classes. A feature class is a group of points, lines, or polygons representing geographic objects of the same kind, like countries or rivers.

A shapefile is a single feature class. Cities.shp, for example, might be a point feature class representing cities. Geodatabases, by contrast, are sets of feature classes. World.mdb might comprise a polygon feature class of world countries, a polyline feature class of world rivers, a point feature class of world cities, and more. (Coverages and CAD files are also sets of feature classes, though in a different way. In a coverage of countries, a polygon feature class would represent countries as polygons, while an arc feature class would represent the same countries as lines.)

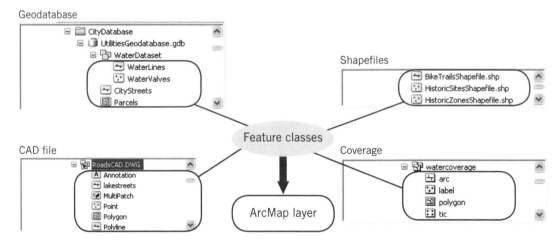

One of the main reasons for the variety of data formats has to do with the rapid changes in GIS technology over the past several years. Shapefiles and coverages, for example, are associated with stages of ESRI software development. While there are differences among spatial data formats, there is also a great deal of mutual compatibility and even interconvertibility. When you add spatial data to ArcMap, you see it and work with it as a layer, and usually you aren't interested in the format of the source file.

You should know, however, that geodatabases, as the latest and most sophisticated spatial data format, have certain advantages. One of these is the ability to store multiple feature classes. Another, mentioned in chapter 7, is the ability to store labels as annotation. A third is the ability to create domains for attributes. A domain establishes valid values or ranges of values for an attribute field and minimizes data entry mistakes by prohibiting invalid values. For example, if "Open" and "Closed" were the domain values for a water valve status field, it would be impossible to enter other values in the attribute table.

Geodatabases offer other important benefits, but many of these require an ArcEditor or ArcInfo license and are beyond the scope of this book. Two types of geodatabases are file and personal. In this chapter, you will create and work with a file geodatabase. A file geodatabases is designed to store datasets in a folder of files on your computer. They can store large datasets (up to one terabyte) and can be accessed by several users at the same time, but only edited by one person at a time. Personal geodatabases have been used in ArcGIS since their initial release in version 8.0 and have used the Microsoft Access data file structure (the .mdb file). They support geodatabases that are limited in size to 2 GB or less. However, the effective database size is smaller, somewhere between 250 and 500 MB, before the database performance starts to slow down. Multiuser geodatabases, on the other hand, are suitable for large workgroups or enterprises. They have no size limit, permit simultaneous editing by different users, and work with relational database management systems such as Informix, Microsoft SQL Server, IBM DB2, or Oracle.

For more information about the new file-based geodatabase, click the Contents tab in ArcGIS Desktop Help and navigate to *Professional Library > Data Management > Managing geodatabases > An overview of the geodatabase > Types of geodatabases.*

14
15
16
17

Creating a personal geodatabase

In ArcCatalog, you can create geodatabases, shapefiles, or coverages. (To create coverages, though, you need an ArcEditor or ArcInfo license.) You can also import and export data from one format to another. In this exercise, you will create a personal geodatabase and import coverage and shapefile data into it.

Exercise 14a

You work in the GIS department of a medium-sized city in Kansas. You are part of a team that is deciding whether the city should convert its shapefile and coverage data to geodatabase format. One question that needs to be answered is how difficult the process would be. To find out, you will create a personal geodatabase and import a land parcels coverage, a water valves shapefile, and a fire hydrants shapefile.

1   Start ArcCatalog. In the catalog tree, double-click **C:\ESRIPress\GTKArcGIS**. Double-click **Chapter14**. Right-click MyData, point to New, and click File Geodatabase.

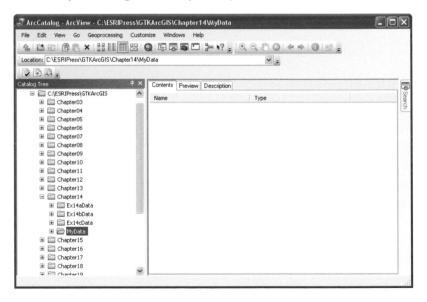

A new personal geodatabase is created in the MyData folder.

You will give the geodatabase a more descriptive name.

2   Make sure the name New Personal Geodatabase.mdb is highlighted inside a black rectangle. (If it isn't, right-click it and click Rename.) Type **CityData** and press Enter.

The geodatabase is now named CityData.gdb.

If you don't see the .gdb file extension, click the Customize menu, click ArcCatalog Options, click the General tab, uncheck Hide file extensions, and click OK. If your CityData icon looks different from the graphic, click the Details button on the Standard toolbar.

14

15

16

17

Next, you'll import the polygon feature class from a land parcels coverage to the new CityData geodatabase.

3   In the ArcCatalog display window, right-click CityData.mdb, point to Import, and click Feature Class (single).

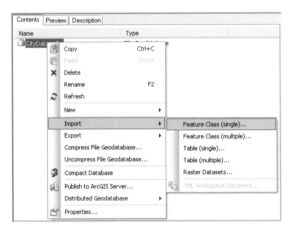

The Feature Class To Feature Class tool opens.

4   Move the dialog box away from ArcCatalog. In the catalog tree, double-click **Ex14aData**, then click the plus sign next to the **parcels** coverage. Click the polygon feature class and drag it to the Input Features box of the dialog box.

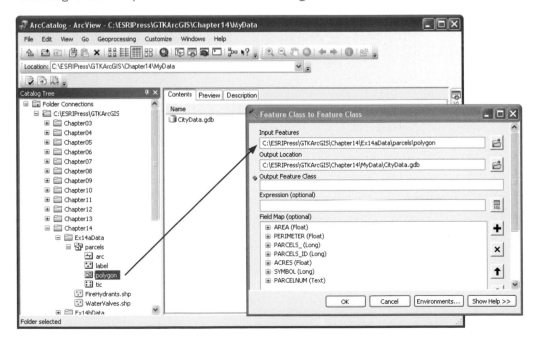

The path to the feature class displays in the Input Features box. In the Output Location box, ArcCatalog correctly assumes that you want to import the data to the CityData geodatabase. (You could change the output location if you wanted to.)

5 In the Output Feature Class box, type **Parcels**. Make sure your dialog box matches the following graphic, then click OK.

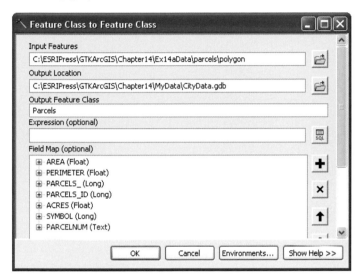

When the operation is completed, a pop-up notification appears by the system task bar. The coverage polygons are converted to a feature class in the CityData geodatabase.

6 In the catalog tree, expand the MyData folder, if necessary, and double-click **CityData.gdb** to see the new feature class. (If you don't see it, click the View menu and click Refresh.)

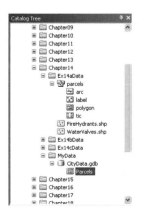

14 15 16 17

Now you will preview the data to make sure it looks okay.

7   In the catalog tree, click the **Parcels** feature class. In the display window, click the
Preview tab.

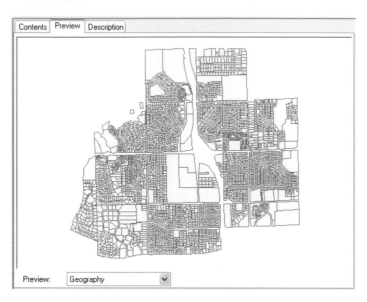

You still have two shapefiles to import into your geodatabase.

8   In the catalog tree, right-click **CityData.gdb**, point to Import, and click Feature
Class (multiple).

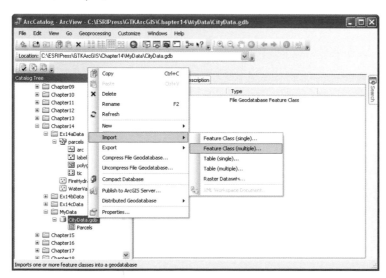

The Feature Class to Geodatabase (multiple) tool opens.

9   Move the dialog box away from ArcCatalog. In the catalog tree, make sure the Ex14aData folder is expanded. Click **WaterValves.shp** and drag it to the Input Features box of the dialog box.

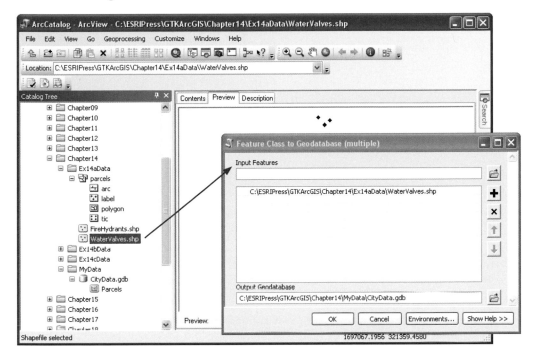

When you drop WaterValves.shp, its path displays beneath the Input Features box in the first row of a list that can hold multiple feature classes.

10   In the catalog tree, click **FireHydrants.shp** and drag it to the Input Features box.

The path to the fire hydrants shapefile is added to the list. The Output Geodatabase box is already correctly set to the CityData geodatabase.

14

15

16

17

11 Make sure your dialog box matches the following graphic, then click OK.

When the operation is completed, a pop-up notification appears by the system task bar.

12 In the catalog tree, expand the MyData folder, if necessary, and double-click **CityData.gdb** to see the new feature classes. Click the **FireHydrants** feature class to preview it, then click **WaterValves** to preview it as well.

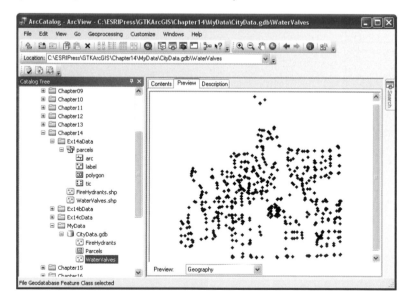

13 In the display window, click the Contents tab.

14 If you are continuing to the next exercise, leave ArcCatalog open. Otherwise exit the application.

Creating feature classes

When you imported data in the previous exercise, you accepted the spatial properties of the existing data. When you create a new feature class, you have to define these properties yourself. This means specifying the feature class geometry—point, line, or polygon—and the spatial reference.

A spatial reference is made up of a coordinate system, a spatial domain, and a precision. As you learned in chapter 13, a coordinate system is a framework for locating features on the earth's surface using either latitude–longitude or x,y values. A spatial domain defines the bounding coordinates for a feature class, beyond which features cannot be stored. Precision defines the smallest measurement that can be made in the coordinate system. For example, if the coordinate system units are feet, a precision of 12 allows you to make measurements as small as an inch. If the coordinate system units are kilometers, a precision of 1,000 allows you to make measurements as small as a meter.

The task of defining a spatial reference is simplified by the fact that you can select from a list of coordinate systems. ArcCatalog will set default domain and precision values for you. You can also import the spatial reference from another dataset and modify it or use it as is.

Exercise 14b

You now know that you can convert the city's existing spatial data to a geodatabase. You also need to know how to create new data in the geodatabase format. In this exercise, you will make a feature class for water lines, although you won't put any features in it until the next chapter.

1   If you have completed exercise 14a, skip to step 2. Otherwise, start ArcCatalog. In the catalog tree, double-click **C:\ESRIPress\GTKArcGIS**. Double-click **Chapter14**. Double-click **Ex14bData**. Right-click **CityData.gdb** and click Copy. Right-click **MyData** and click Paste.

14
15
16
17

2 Start ArcCatalog, if necessary. In the catalog tree, double-click **C:\ESRIPress\GTKArcGIS**. Double-click **Chapter14**, then double-click **MyData**. Right-click the CityData file geodatabase, point to New, then click Feature Class.

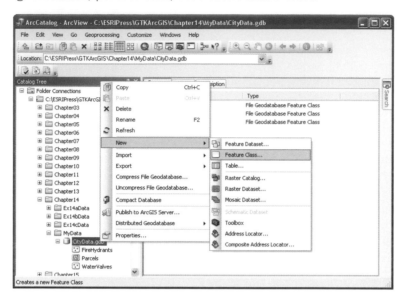

The first panel of the New Feature Class wizard opens.

3 In the Name box, type **WaterLines**.

The default Type is Polygon Features. Since you are creating a feature class of water lines, you'll change this to Line Features.

4 Click the Type drop-down arrow and click Line Features.

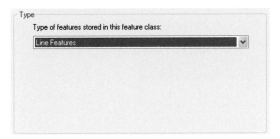

5 Click Next.

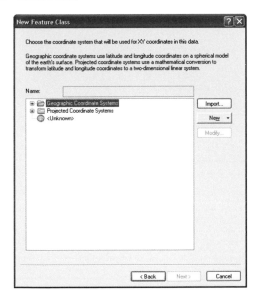

As no coordinate system has been selected, the Name is unknown and the Name box is empty.

6 Click Import.

14

15

16

17

7   In the Browse for Coordinate System dialog box, navigate to **\GTKArcGIS\Chapter14 \MyData** (if you didn't complete exercise 14a, navigate to **\GTKArcGIS\Chapter14 \Ex14bData.**) Double-click **CityData.gdb** and click **Parcels**. Make sure that the dialog box matches the following graphic, then click Add.

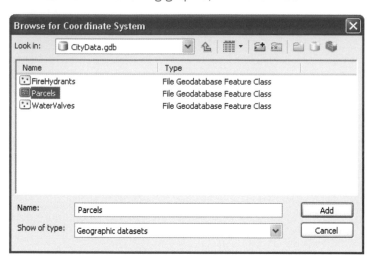

The name of the coordinate system you imported appears in the Name box.

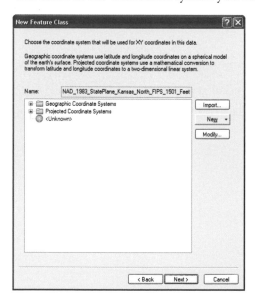

8 Click Next.

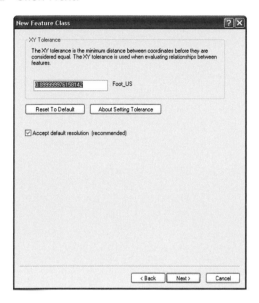

The XY tolerance defines the minimum distance between two coordinates before they are considered equal. You will accept the default value.

9 Click Next.

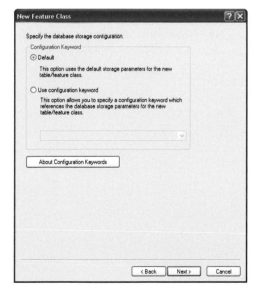

A database storage configuration enables you to fine tune how data is stored in a file geodatabase. You will accept the default option.

14
15
16
17

10 Click Next.

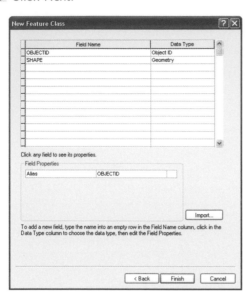

The fifth panel of the wizard displays the field names, data types, and field properties that belong to the new feature class. OBJECTID and SHAPE are required fields, automatically added by the software. The OBJECTID field stores a unique ID number for every feature in the class. The SHAPE field stores each feature's shape and location.

Depending on the type of feature class created, ArcGIS also creates and maintains measurement fields. A SHAPE_Length field is created for line feature classes. A SHAPE_Length and a SHAPE_Area field are created for polygon feature classes. (The SHAPE_Length attribute for polygons stores perimeter lengths.) These measurement fields appear only after the feature class is created.

11 In the Field Name column, click SHAPE. The field properties for the SHAPE field are displayed.

Notice that the geometry type is set to Line.

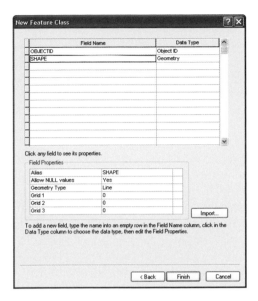

12 Click Finish.

13 In the catalog tree, if necessary, click the plus sign next to **CityData.gdb**.

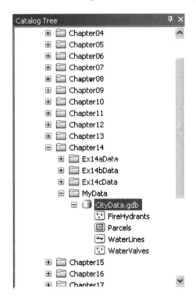

WaterLines has been added to the list of feature classes.

In the next exercise, you will add two new attributes to the WaterLines feature class: one for the date water lines are installed and one for the type of water line.

14 If you are continuing to the next exercise, leave ArcCatalog open. Otherwise, exit the application.

Adding fields and domains

When you create a feature class, you add fields to it to store attribute information. A field is defined by a name, a data type (for instance, text or integer), and properties that vary according to both the data type and the spatial data format. One such property, which can be defined for fields in geodatabases, is a domain.

A domain is either a list or a range of valid values for an attribute. A city zoning attribute, for example, might have a dozen or so appropriate values, such as "residential," "industrial," and "commercial." By defining these values as a domain, you prevent any other values from being mistakenly added to a table during data entry or editing. Similarly, you might want to restrict the possible values for a numeric attribute to a particular range, such as 0 to 14 for a soil pH attribute.

Unlike other field properties, which apply exclusively to the field they are defined for, a domain can be applied to more than one field in a feature class and to more than one feature class within a geodatabase.

Exercise 14c

In this exercise, you will add two fields to the WaterLines feature class you created previously. One will store installation dates, the other the type of water line. There are three types of water lines: mains (which run under streets), domestic laterals (which run from mains to houses), and hydrant laterals (which run from mains to fire hydrants).

After you add the fields, you will create a domain and apply it to the field for water line type. The domain will ensure that no values other than the three valid ones can be entered in the attribute table. You will then specify one of those three to be the default.

1   If you have completed exercise 14b, skip now to step 2. Otherwise, start ArcCatalog. In the catalog tree, double-click **C:\ESRIPress\GTKArcGIS**. Double-click **Chapter14**. Double-click **Ex14cData**. Right-click **CityData.gdb** and click Copy. Right-click **MyData** and click Paste.

2 Start ArcCatalog, if necessary. In the catalog tree, double-click **C:\ESRIPress\GTKArcGIS**. Double-click **Chapter14**. Double-click **MyData** and double-click **CityData.gdb**.

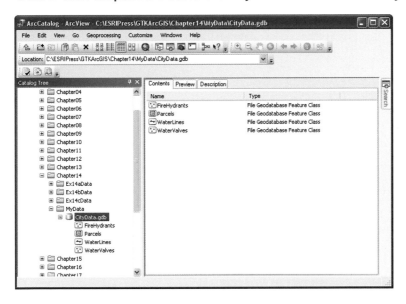

3 In the catalog tree, right-click **WaterLines** and click Properties. In the Feature Class Properties dialog box, click the Fields tab, if necessary.

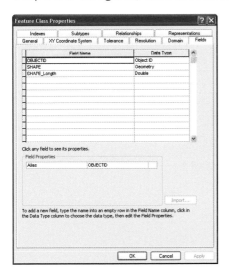

4 In the Field Name column, type **Install_Date** in the first empty row. (Spaces are not allowed in field names.)

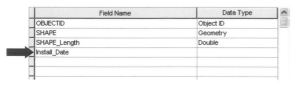

5 In the Data Type column, click the empty cell next to Install_Date. In the drop-down list that displays, click Date.

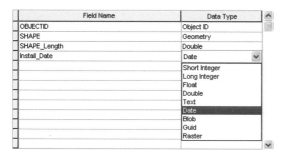

6 In the Field Name column, type **Line_Type** in the next empty row. In the Data Type column, click the empty cell next to Line_Type. In the drop-down list, click Text. Make sure the dialog box matches the following graphic, then click OK.

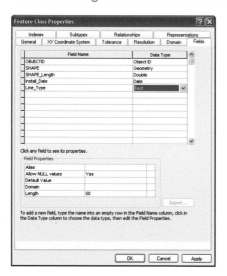

Now that you have added the two attribute fields, you will create a domain and apply it to the Line_Type field.

7   In the catalog tree, right-click **CityData.gdb** and click Properties to display the Database Properties dialog box. Click the Domains tab.

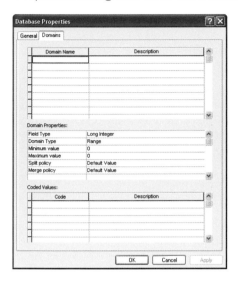

8   In the Domain Name column, type **WaterLineType** in the first cell. Click the first cell in the Description column and type **Type of water line installed**.

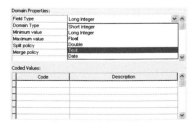

In the Domain Properties list, the default Field Type value is Long Integer and the Domain Type is Range. The field type must match the data type of the feature class it is being applied to. Since you made Line_Type a text field in step 6, you must do the same here.

9   Click the Long Integer value. In the drop-down list, click Text.

14
15
16
17

The domain type changes automatically from Range to Coded Values.

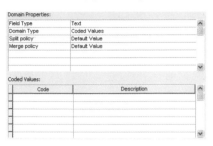

A Coded Values domain is used when the attribute values you want to enforce are best defined as a list. (The values may nonetheless be numeric because numbers are commonly used as codes for descriptions.) A Range domain is used when the values you want to enforce are quantities with upper and lower limits.

10 In the Coded Values list at the bottom of the dialog box, click the first empty cell in the Code column. Type **M**. Click the first cell in the Description column and type **Main**.

11 In the next empty cell in the Code column, type **DL**. Type **Domestic Lateral** as its description.

12 Type **HL** in the next empty Code column cell. Type **Hydrant Lateral** as its description. Make sure the dialog box matches the following graphic, then click OK.

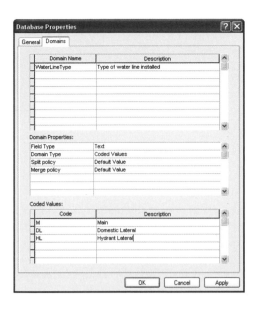

After a Coded Value domain has been applied to a feature class, you will pick from a drop-down list of the descriptions when you enter attribute values. The coded values are stored in the geodatabase, but you only see the descriptions in the attribute table.

Now that you have created the domain, you will apply it to the Line_Type field of the WaterLines feature class.

13 In the catalog tree, right-click the **WaterLines** feature class and click Properties. In the Feature Class Properties dialog box, click the Fields tab if necessary.

14 In the Field Name column, click Line_Type. Its properties display.

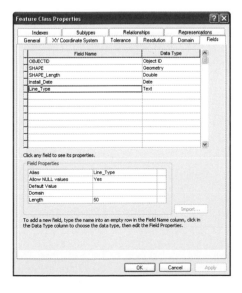

15 In the Field Properties list, click the empty cell next to Domain. In the drop-down list, click WaterLineType.

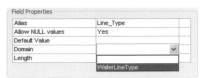

You have created a domain and applied it to the Line_Type field. Finally, you will assign the field a default value. Mains are the most common type of water lines, so you'll make "Main" the default. Whenever a water line is added to the feature class, it will automatically get this value in the Line_Type field. As you'll see in the next chapter, you can easily change the value if the line is a lateral.

16 In the Field Properties frame, click the empty cell next to Default Value and type **M**, the code for mains. Make sure the dialog box matches the following graphic, then click OK.

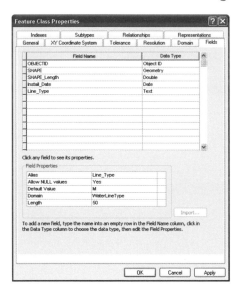

You can also set a default value for a field without creating a domain. If a domain is applied, however, the default value must be a valid domain value.

17 Close ArcCatalog.

Chapter 15

Creating features

Drawing features
Using feature construction tools

Much of the vector spatial data used in a GIS has been digitized from paper maps and aerial or satellite photographs. Digitizing data involves placing a map or photo on a digitizing tablet (a drawing table connected to a computer) and tracing features with a puck, which is a device similar to a mouse. In a variation called heads-up digitizing, features are drawn with a mouse directly on the computer screen by tracing an aerial photo, a scanned map, or other spatial data.

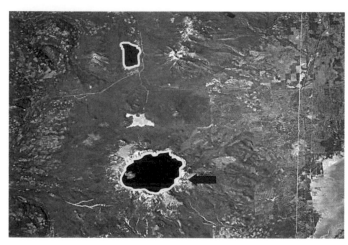

Oregon's Crater Lake digitized heads-up from a satellite image.

Features can also be digitized without tracing. ArcGIS has several tools for creating circles, rectangles, curves, and other shapes of exact dimensions.

Points are the simplest features to digitize—all it takes is a single click. Lines are more complicated because they start, end, and may change direction. Polygons, from the point of view of digitizing, are just special cases of lines—they are lines that return to their origins.

The places where a line begins and ends are called endpoints. The places where a line changes direction or is intersected by another line are called vertices. The segments between vertices are called edges.

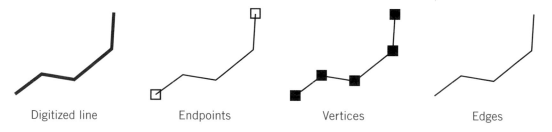

Digitized line   Endpoints   Vertices   Edges

When you look at line or polygon features on a map, you don't see them as edges, vertices, and endpoints, but when you create or edit features, you may. When you see a feature in this way, you are looking at its edit sketch. A new feature exists only as an edit sketch until it is saved.

In ArcGIS, all digitizing—whether to create new features or modify existing ones—is done in the course of an edit session. An edit session begins when you click the Start Editing menu command on the Editor toolbar and ends when you click the Stop Editing command. After starting an edit session, creating features is accomplished through the use of feature templates. Feature templates define all the information required to create a feature: the layer where a feature will be stored, the attributes a feature is created with, and the default tool used to create that feature.

Editor menu   Create features

The Editor toolbar

A window, called the Create Features window, is the central place to create and manage feature templates. The window has three main components: a toolbar to manage your templates and their properties, a list of templates used to create new features, and a set of tools used to define the features' shape. You'll learn how to use this window and its options in this chapter.

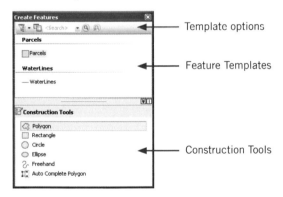

Template options

Feature Templates

Construction Tools

When you digitize features, you sometimes make mistakes. You can undo mistakes with the Undo button on the Standard toolbar.

Drawing features

In this exercise, you will create water lines for the feature class you made in the CityData geodatabase in exercise 14b.

New features are often connected to existing features. For example, the boundaries of new land parcels may be adjacent to boundaries that have already been digitized. New streets intersect or extend to existing ones. New power lines run to and from poles. Making sure that features are connected would be difficult if it depended entirely on your eyesight.

ArcMap can automatically connect (or snap) features placed within a certain distance of each other. The rules specifying which features, and which feature parts, snap to others make up the snapping environment. The distance at which snapping occurs is called the snapping tolerance.

Exercise 15a

In a subdivision of your growing city, a contractor has laid new water lines, including an extension to a main and a line connecting a main to a fire hydrant. During construction, the city's project inspector recorded the locations of water valves and hydrants with a Global Positioning System (GPS) device. These locations have been imported as point feature classes in your geodatabase. As the city's GIS technician, you will use this point data and the ArcMap editing tools to digitize the new water lines.

1   Start ArcMap. In the ArcMap—Getting Started dialog box, under the Existing Maps section, click Browse for more. (If ArcMap is already running, click the File menu and click Open.) Navigate to **C:\ESRIPress\GTKArcGIS\Chapter15**. Click **ex15a.mxd** and click Open.

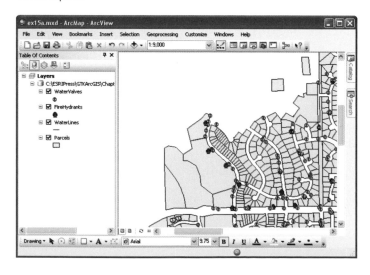

The map shows a subdivision under construction. Layers for water valves, fire hydrants, water lines, and parcels are listed in the table of contents. The WaterLines feature class you created in exercise 14b has no features in it yet. Before starting the edit session, you will zoom to the area where you are going to digitize water lines.

2 Click the Bookmarks menu and click New Water Lines.

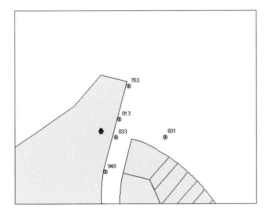

The display zooms to an area where there are several water valves and a fire hydrant. The field crew has laid three new water lines: from valve 763 to valve 813, from 813 to 831, and from 813 to the nearest fire hydrant. You will digitize these three lines.

First, you'll add the Editor toolbar and start an edit session.

3 On the Standard toolbar, click the Editor Toolbar button.

The Editor toolbar opens. The tools are disabled.

4 If necessary, move the Editor toolbar away from the map display. On the Editor toolbar, click the Editor menu and click Start Editing.

The editing tools are now enabled. The Create Features window appears (it may appear docked, tabbed, or floating depending on how you arranged your windows.) By default, no feature template is selected. You need to change the target to the water lines layer.

5   In the Create Features window, click the WaterLines template.

Because the water lines must connect (snap) to each other and to other features, such as valves and hydrants, you'll set the snapping environment before you start digitizing.

6   On the Editor toolbar, click the Editor menu, point to Snapping and click Snapping Toolbar. The Snapping Toolbar opens.

Setting the snapping environment

The Snapping toolbar contains buttons of the main snap types, although additional ones are available on the Snapping menu. As you digitize a new feature, the cursor snaps to existing features according to which boxes are enabled. (Enabling either the Vertex Snapping or Edge Snapping for a point layer makes the cursor snap to point features.) The cursor snaps when it comes within a specified distance of a feature. To set this snapping tolerance, click the Snapping menu in the Snapping toolbar, then click Options.

The Snap To Sketch option in the Snapping menu sets snapping rules that an edit sketch uses on itself. For example, it is useful when you are creating a feature that intersects or connects to itself, such as a boundary line that forms a closed loop. For more information, click the Contents tab in ArcGIS Desktop Help and navigate to *Professional Library > Data Management > Editing data > Using snapping > About snapping.*

7 Click the End, Vertex, and Edge Snapping buttons to disable them. Only the Point Snapping button is left enabled.

You need to digitize the water lines.

8 In the Create Features window, in the Construction Tools panel, click Line.

9 Move the mouse pointer over the map display. The cursor changes to a crosshair. Move the cursor toward valve 763.

When you come within the snapping tolerance of the valve, a circle appears and a "WaterValve: Point" identifies it.

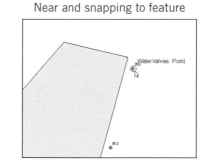

Now you will digitize the first water line, from valve 763 to valve 813. To draw a line feature, you click once to begin the line, click again for each vertex, and double-click to finish the line. Because the water lines don't change direction, you won't need any vertices.

10 With the cursor snapped to valve 763, click to start the first water line. Move the cursor toward valve 813. The edit sketch of the line moves with the mouse (an effect known as rubberbanding).

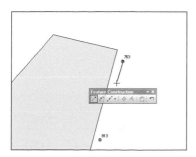

The Feature Construction toolbar appears nearby the cursor (for convenience). Leave it set to Straight Segment.

14
15
16
17

11 Move the cursor over valve 813. When the cursor snaps to it, double-click to end the line. (If you make a mistake, click the Undo button on the Standard toolbar and repeat steps 10 through 12.)

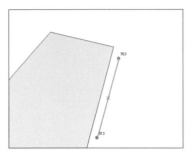

The line is created. Now you will add a second line connecting valve 813 to valve 831.

12 With the cursor still snapped to valve 813, click to begin a new line. Move the cursor over valve 831. When the cursor snaps to it, double-click to end the line.

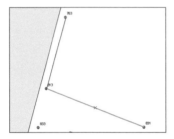

The first line changes from cyan (the default color of edit sketches) to Cretean blue (the color symbolizing water lines in the map). Now, you will digitize the third and final water line, from valve 813 to the fire hydrant.

13 Move the cursor over valve 813 so that the cursor snaps to it. Click to start the line. Move the cursor over the fire hydrant. When the cursor snaps to it, double-click to end the line.

Now you will view the attributes for individual water lines that you just created.

14 With the water line to the fire hydrant still selected, on the Editor toolbar, click the Attributes button.

The Attributes window opens.

The Attribute window can show one or more features at a time. Use the Edit tool on the Editor toolbar to select features to see their attributes. Alternatively, open the attribute table of the WaterLines layer in the table of contents to view all of the records without the need to select them.

The length of each line, in feet, has been calculated in the SHAPE_Length field. (Length and area attributes maintained by the software accurately measure features in the geodatabase, but they do not have legal authority. An official utility database would include an "as-built" length attribute containing the results of field surveys.)

The Install_Date field is <Null> because no attribute values have been entered for it. The Line_Type attribute values have been set to Main for all three water lines, in accordance with the default value for water lines you specified in exercise 14c. The default value isn't right for the last line you digitized, which connects a valve to a hydrant.

14
15
16
17

15 In the Line_Type field, click Main. In the drop-down list that appears, click Hydrant Lateral.

You are finished digitizing for now and can save your work.

16 Close the attribute table. On the Editor toolbar, click the Editor menu and click Stop Editing. Click Yes to save your edits. Close the Snapping toolbar. Close the Editor toolbar. (If you docked a toolbar, you must undock it, then close it.)

There is no need to save the map document. Your edits were saved to the feature class in the geodatabase. Any map document that contains the WaterLines layer will include the new features.

17 If you are continuing to the next exercise, leave ArcMap open. Otherwise, exit the application. Click No when prompted to save your changes.

Using feature construction tools

In this exercise, you will digitize features representing land parcels. The basic procedure—starting an edit session and setting the snapping environment—is the same as in the previous exercise, but you will use some different tools. ArcMap can set the lengths and angles of edit sketch edges to precise specifications. It also has tools for automatically completing some operations (like finishing a square) once you provide the necessary information.

Exercise 15b

You have left your job with the city and gone to work for a developer who is planning to subdivide a large tract of land. Having obtained parcel data from the city's GIS department, you will use preliminary field measurements to digitize some new parcels in accordance with the developer's proposal. Eventually, your map will go to the city's planning department for review. If the design is approved, the developer will commission a licensed survey crew to conduct a legal survey of the parcel boundaries.

1 In ArcMap, open **ex15b.mxd** from the **C:\ESRIPress\GTKArcGIS\Chapter15** folder.

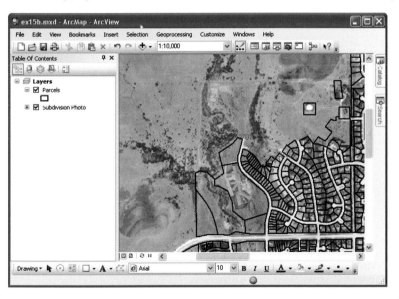

The map shows the parcels displayed against a background photo that lets you see what has been built on the parcels. Some have finished houses, some have houses under construction, and some are vacant.

14
15
16
17

You will zoom in and digitize a new parcel next to parcels 2731 and 2726.

2 Click the Bookmarks menu and click First Parcel Site.

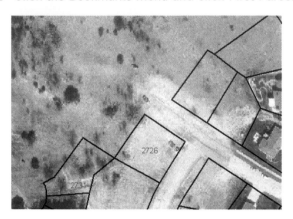

3 On the Standard toolbar, click the Editor Toolbar button.

4 On the Editor toolbar, click the Editor menu and click Start Editing.

First, you will set the snapping setting so that boundary lines for the new parcels snap to the vertices of existing parcels. In the Create Features window, the Parcel feature template is chosen since there is only one feature in the map document.

5 On the Editor toolbar, click the Editor menu, point to Snapping, and click Snapping Toolbar. In the Snapping toolbar, check the Vertex button to enable it as shown in the following graphic. Close the Snapping toolbar.

The new parcel will share two sides with existing parcels, as shown in the following graphic. A task called Auto Complete Polygon is used to digitize features that share boundaries with existing features. You will digitize the two new sides and ArcMap will finish the polygon.

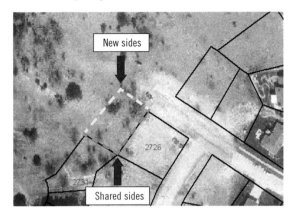

6 In the Create Features window, click Parcels, and then click Auto Complete Polygon in the Construction Tools panel.

7 In the table of contents, turn off the Subdivision Photo layer to see the parcels more clearly.

8   Move the cursor over the northern most corner of parcel 2731 so that the cursor snaps to its vertex (the label Parcels: Vertex appears). Click to begin the polygon. (The Feature Construction toolbar appears nearby, you may move it out of the way if you wish.)

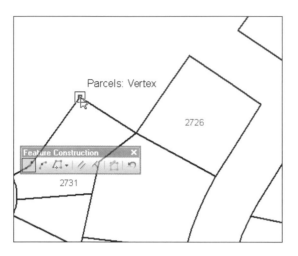

The two boundaries you'll digitize must be aligned with the boundaries of the existing parcels. Your field notes say that the new parcel's northwest boundary should be parallel to the northwest boundary of parcel 2731 and that it should be 118 feet long.

9   Right-click anywhere on the northwest boundary of parcel 2731 to open the context menu. On the context menu, click Parallel. (If you left-click by mistake and add a vertex, click the Undo button on the Standard toolbar.)

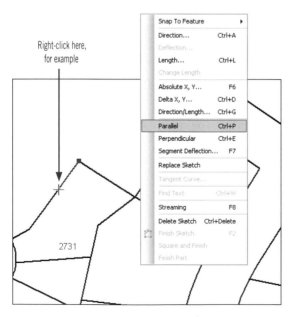

You can use the Escape key on the keyboard to close the context menu without choosing a menu command. The Escape key will also cancel the effect of a menu command.

10 Move the cursor over the map display without clicking. The line is constrained to be parallel to the line segment you just clicked.

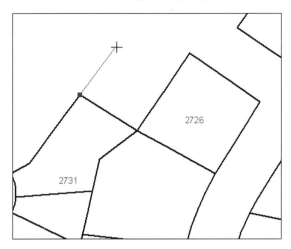

11 With the cursor located anywhere in the display, right-click. On the context menu, click Length. In the Length pop-up box, delete the existing value and type **118** as shown in the following graphic, then press Enter. (The graphic shows both the context menu and the pop-up box; in reality, the menu disappears when the box displays.)

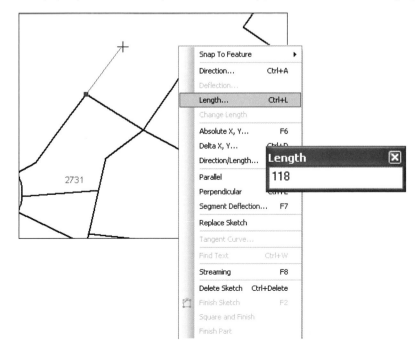

The first side of the new parcel is finished. It is 118 feet long and runs parallel to the boundary of parcel 2731. The cursor moves to anticipate the next line segment. (Yours may move to a different position from the one in the graphic.)

12 Move the cursor over the northern most corner of parcel 2726 so that the cursor snaps to its vertex. Double-click to end the line segment. If you moved your cursor away from the vertex, right-click anywhere and click Finish Sketch in the context menu.

The third and fourth sides of the polygon are completed automatically, using the existing parcel boundaries.

Now you will add a second parcel to a different part of the subdivision.

13 In the table of contents, turn on the Subdivision Photo layer. Click the Bookmarks menu and click Second Parcel Site. You will digitize the new parcel in the area shown by the yellow box below. (The box appears only in the book, not on your screen.)

This time you don't have existing boundaries to work from, so you'll change the task to Polygon in the Construction Tools panel. Your notes tell you that the new parcel should have a corner at a point exactly 61.1 feet from a corner of parcel 2719 and 119.4 feet from a corner of parcel 2707. You will use the Distance-Distance tool, on the Sketch tool drop-down palette to choose the point of beginning for the new parcel.

14 In the Create Features window, click Parcels, and then click Polygon in the Construction Tools panel.

15 Click the Sketch tool drop-down arrow and click the Distance-Distance tool.

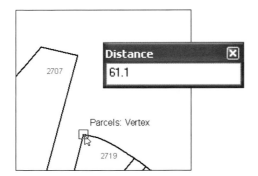

16 In the table of contents, turn off the Subdivision Photo layer.

17 Move the cursor over the northwest corner of parcel 2719 and click when the cursor snaps to its vertex. Press the D key on the keyboard once. In the Distance pop-up box, type **61.1**, as shown in the following graphic, then press Enter.

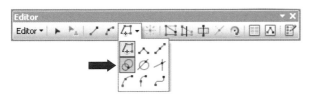

A circle appears on the map. Its center is the vertex you clicked; its radius is 61.1 feet.

18 Move the cursor over the northeast corner of parcel 2707 and click when the cursor snaps to its vertex. Press the D key on the keyboard once. In the Distance pop-up box, type **119.4**, as shown in the following graphic, then press Enter.

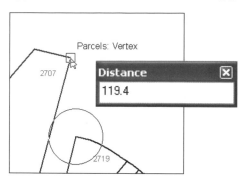

A second circle is drawn, with a radius of 119.4 feet. The circles intersect at two points.

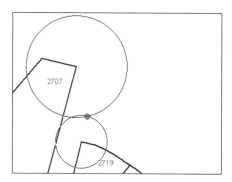

19 In the area of intersection, move the cursor left and right without clicking.

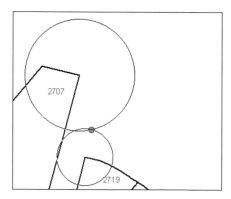

The highlighted dot is constrained to move from one point of intersection to the other. Theoretically, either could be the correct point of beginning. It so happens, however, that the point on the left would start the parcel in the middle of an unbuilt road. The point on the right is the one you want.

20 Move the cursor to the point on the right and click. The circles disappear. Move the cursor slightly away. A red square marks the spot. The Feature Construction toolbar appears nearby.

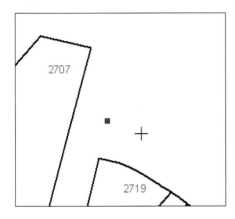

You are now ready to digitize the parcel, using field notes detailing the length and angle of the parcel sides.

21 On the Feature Construction toolbar, click the Straight Segment tool.

This tool is the same as the one on the Editor toolbar.

14

15

16

17

22 Right-click anywhere on the map display. On the context menu, click Direction. In the pop-up box, replace the current value with **73.4**, as shown in the following graphic, then press Enter.

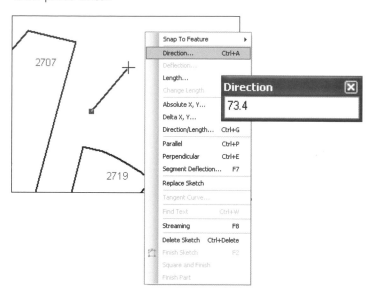

The line is constrained to the angular direction you set. The Direction command uses east as 0 degrees and measures positive angles counterclockwise.

Note: If your Direction pop-up box looks different from the one shown in step 22, your Direction Type may be set incorrectly. To change the setting, on the Editor toolbar, click the Editor menu and click Options. In the Editing Options dialog box, click the Units tab. Change the Direction Type to Polar. Make sure your settings match those shown in the following graphic.

23 Right-click anywhere above the starting point. On the context menu, click Length. In the pop-up box, replace the current value with **54.87**, as shown in the following graphic, then press Enter.

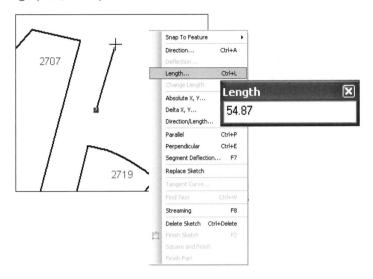

The first side of the parcel is complete. The starting point changes to a green square and the current vertex becomes a red square.

You will now set a direction and a length to create the second line segment.

24 Right-click anywhere to the right of the red vertex. On the context menu, click Direction. In the pop-up box, replace the current value with **-15.72** (note the minus sign), as shown in the following graphic, then press Enter.

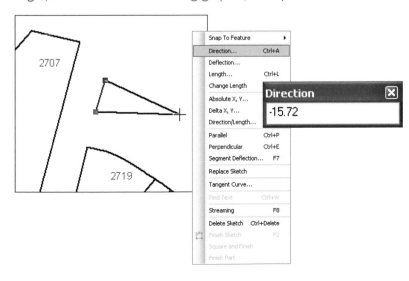

14
15
16
17

25 Right-click anywhere to the right of the red vertex. On the context menu, click Length. In the pop-up box, replace the current value with **117**, as shown in the following graphic, then press Enter.

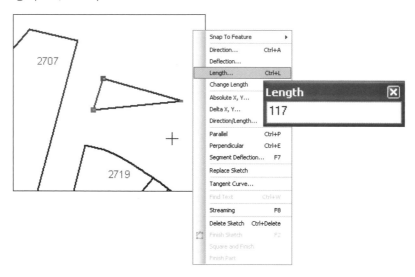

The second side of the parcel is done. To complete the last two sides, you'll use the Square and Finish command, which completes a polygon by drawing a right angle.

26 Right-click anywhere below the red vertex. On the context menu, click Square and Finish.

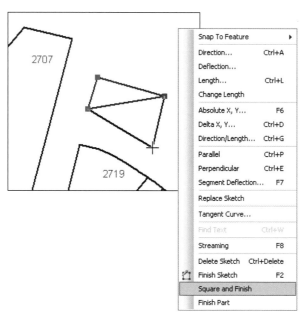

The parcel is drawn to your specifications.

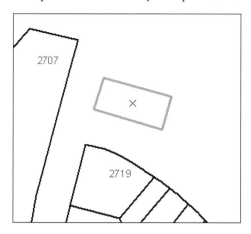

27 On the Editor toolbar, click the Editor menu and click Stop Editing. Click Yes to save your edits. In the table of contents, turn on the Subdivision Photo layer.

You have added two proposed parcels to the subdivision. The developer has more in mind, but you won't have to digitize them in this book. You don't need to save the map document. Your edits were made directly to the feature class in the geodatabase. When you saved them, the geodatabase was updated.

28 Close the Editor toolbar. If you are continuing to the next chapter, leave ArcMap open. Otherwise, exit the application. Click No when prompted to save your changes.

14
15
16
17

Chapter 16

Editing features and attributes

Deleting and modifying features
Splitting and merging features
Editing feature attribute values

Creating features is one part of maintaining a geodatabase. Another part, equally important, is making changes to existing features.

Features can be edited in several ways. They can be deleted, moved, split, merged, resized, reshaped, or buffered. In addition to the tasks on the Editor toolbar, there are commands on the Editor menu for changing features.

When features are edited, their attributes often need to be updated as well. In a geodatabase, length and area attributes are updated automatically by ArcMap; other attributes must be updated manually. Sometimes attribute values need to be edited even when features don't change—for example, when a land parcel is sold, the owner name needs to be changed. And sometimes new attributes are added to a table.

With an ArcView license, you can edit file and personal geodatabases and shapefiles. With an ArcEditor or ArcInfo license, you can also edit coverages and multiuser geodatabases. Files in CAD format cannot be edited directly, but can be converted to an editable format. To edit raster layers, you need the ArcGIS Spatial Analyst extension.

Deleting and modifying features

To modify a feature, you add, delete, or move vertices on the feature's edit sketch. If a feature shares an edge or vertex with another feature, you may want to ensure that no gap or overlap results. You can make touching features behave as if they are connected by creating a map topology within your edit session.

Exercise 16a

You are the GIS specialist for the city's planning department. To prepare for a meeting of the City Commission, you will make some changes proposed for parcels in a new subdivision. Since the changes are recommendations only, you will not work with the city's live geodatabase, but with a copy designated for planning use.

Your first task is to delete two parcels on land the city is negotiating to buy back from the subdivision developer. Your second task is to enlarge the boundaries of two parcels to make room for a pond to catch water runoff from a nearby road and sidewalk. The map is simply for discussion, so your edits don't have to be exact. If the proposals are implemented, a licensed survey crew will resurvey the parcels.

1   Start ArcMap. In the ArcMap—Getting Started dialog box, under the Existing Maps section, click Browse for more. (If ArcMap is already running, click the File menu and click Open.) Navigate to **C:\ESRIPress\GTKArcGIS\Chapter16**. Click **ex16a.mxd** and click Open.

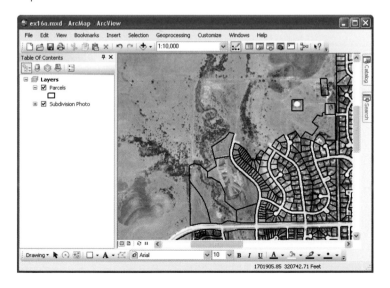

The parcels are outlined in black against an aerial photo of the subdivision. You will zoom in on the parcels to be deleted.

2   Click the Bookmarks menu and click First Parcel Site.

The parcels are labeled with their ID numbers. (The labels are set to display at scales of 1:4,000 and larger.) You will delete parcels 2712 and 2718 in the middle of the display.

3   On the Standard toolbar, click the Editor Toolbar button.

4   On the Editor toolbar, click the Editor menu and click Start Editing.

The target layer is Parcels, the only editable layer in the map. The task is set to Create New Feature. You can delete selected features regardless of which task is set.

5   On the map, click anywhere on parcel 2712. It is outlined in cyan to show that it is selected. Hold down the Shift key and click parcel 2718 to select it also.

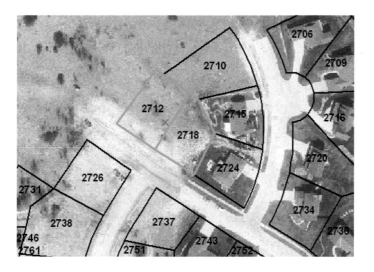

The small blue "x" in the center of the selected parcels is called a selection anchor. Every selection in an edit session, whether of one feature or many, has one. (Selection anchors are used to rotate and scale features, which you won't do in this exercise.)

6   With parcels 2712 and 2718 selected, click the Delete button on the Standard toolbar.

The selected features are deleted.

7   On the Editor toolbar, click the Editor menu and click Save Edits.

Now you will modify the shapes of two parcels to accommodate the proposed detention pond. You need to move the parcel edges to a nearby sidewalk. You will add a vertex to one parcel's edit sketch and move another vertex that is shared by two parcels.

8   Click the Bookmarks menu and click Second Parcel Site.

The proposed pond is outlined in yellow in the book (but not on your screen).

The parcels to be edited are 2768 and 2845. The parcels will be extended to the sidewalk at the bottom of the display. The sidewalk appears on the image as a thin white line that runs east—west and lies just north of the road. To make the edits, you will zoom in closer.

9 Click the Bookmarks menu and click Proposed Drainage Change.

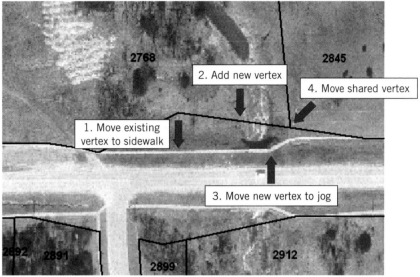

You will make four edits. First, you'll move an existing vertex in parcel 2768 to the sidewalk. Second, you'll add a new vertex to parcel 2768. Third, you'll move the new vertex to the point where the sidewalk jogs. Fourth, you'll move a vertex shared by parcels 2768 and 2845.

10 On the Editor toolbar, click the Edit tool if necessary.

14

15

16

17

11 Double-click anywhere on parcel 2768 to select it and display its edit sketch.

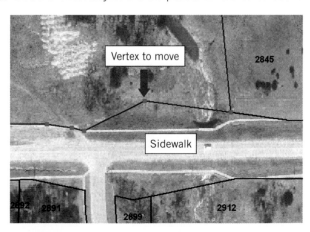

On the edit sketch, the vertices are green squares. (Only the last digitized vertex—which you don't see because you are zoomed in on a corner of the parcel—is a red square.) On the Editor toolbar, ArcMap has automatically enabled the Edit Vertices toolbar.

12 Place the cursor over the first vertex to be moved. The cursor changes to a four-headed arrow.

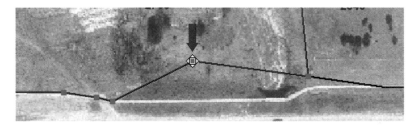

13 Drag the vertex down to the sidewalk and release the mouse button.

The green line shows the new shape of the feature. The highlighted cyan line remains in place as a reference. It will disappear when you unselect the polygon or select another feature.

Now you'll add a new vertex and move it to accommodate the jog in the sidewalk.

14 On the Edit Vertices toolbar, click Add Vertex.

15 Click the black line about where the arrow is in the following graphic.

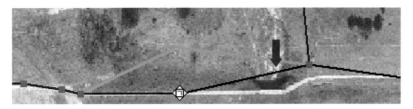

A new vertex is added where you clicked.

16 Place the cursor over the new vertex and drag the vertex to where the sidewalk changes direction. Release the mouse button.

17 Click somewhere outside the polygon to complete the edit.

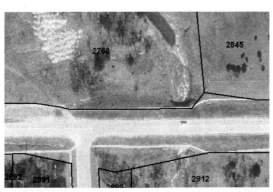

14
15
16
17

18 On the Editor toolbar, click the Editor menu and click Save Edits.

Now you'll move a vertex shared by parcels 2768 and 2845 to the sidewalk. This "shared" vertex is actually two vertices (one belonging to each polygon) that lie on exactly the same spot.

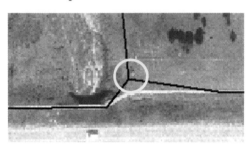

You could use the Edit tool to move each vertex separately. Since you want the parcels to remain adjacent, this would mean setting the snapping environment, moving one parcel's vertex, then moving the second parcel's vertex and snapping it to the first. That wouldn't be too hard this time, but if you had to do it several times, or if you had four or five polygons sharing a vertex, it might start to get inconvenient.

A better approach is to create a map topology. Map topology makes features aware of their connectedness to other features, so that common vertices and edges can be moved together.

19 Click the Customize menu, point to Toolbars, and click Topology. (You may need to scroll down in the list of toolbars.)

The Topology toolbar opens. Only one of its tools is enabled.

20 On the Topology toolbar, click the Map Topology button.

The Map Topology dialog box opens.

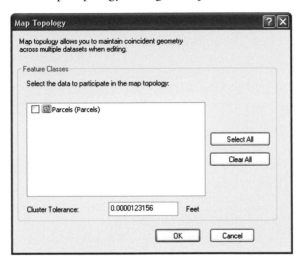

You can choose the layers for which you want to create map topology (in this case, Parcels is the only possibility). You can also set a cluster tolerance, which is the distance at which ArcMap considers nearby edges or vertices to be coincident, or shared.

21 On the Map Topology dialog box, check the Parcels check box and leave the cluster tolerance at its default value. Make sure your dialog box matches the following graphic, then click OK.

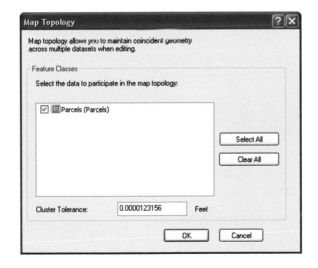

On the Topology toolbar, the Topology Edit tool is now enabled.

14
15
16
17

22 On the Topology toolbar, click the Topology Edit Tool.

23 Place the cursor over the map display and click the location of the shared vertex. The vertex highlights in purple. (If you highlight a polygon edge by mistake, click outside the parcel and try again.)

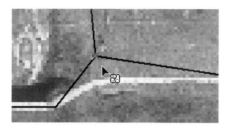

24 Place the cursor over the purple vertex. The cursor changes to a four-headed arrow.

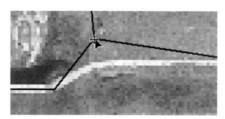

25 Drag the purple vertex to the point where the sidewalk changes direction again. As you drag, both polygons move together. Release the mouse button.

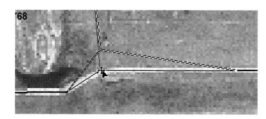
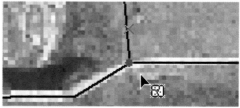

While dragging the vertex   After releasing the mouse button

26 Close the Topology toolbar.

27 On the Editor toolbar, click the Editor menu and click Stop Editing. Click Yes when prompted to save your edits. Close the Editor toolbar.

The map is ready for the City Commission meeting. You don't need to save the map document because your edits were made directly to the feature class in the geodatabase.

Map topology and geodatabase topology

Map topology can be created for shapefiles and geodatabase feature classes. It lasts only for the duration of an edit session and allows you to edit features (within the same layer) as if they were connected to each other. With an ArcEditor or ArcInfo license, you can create a more sophisticated topology for geodatabase feature classes. Geodatabase topology lets you enforce some twenty-five spatial rules within and across feature classes, helping to guarantee the integrity of your data. For example, you can specify that polygons may not overlap or have gaps, that lines must be connected to other lines, or that points in one feature class must lie within polygons of another feature class. Geodatabase topology is stored within the geodatabase and is thus independent of a particular map document or edit session. For more information, click the Contents tab in ArcGIS Desktop Help and navigate to *Professional Library > Data Management > Editing data > Editing topology > Editing shared geometry > About creating a map topology.*

28 If you are continuing to the next exercise, leave ArcMap open. Otherwise, exit the application. Click No when prompted to save your changes.

14

15

16

17

Splitting and merging features

To split a polygon or a line (create two features from one), you digitize a line across the feature where you want to split it. You can split several features at once.

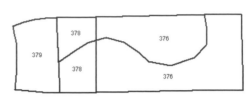

Select polygons to split and draw a split line.

Each selected polygon is split in two.

To merge polygons or lines (create one feature from two or more), you select them and click the Merge command on the Editor menu. The merged features do not have to be contiguous. For example, you could merge a large polygon representing a country and smaller polygons representing its offshore islands. The country and islands would then exist as a multipart feature with a single record in the attribute table.

Merge resembles the Dissolve tool that you used in chapter 11, but whereas Dissolve combines polygons that have the same attribute value, Merge combines any selected polygons, regardless of their attribute values.

Select polygons to merge.

The three polygons are merged into one.

When features in geodatabase feature classes are split or merged, ArcMap updates their Shape_Length and Shape_Area attributes. This is not done for features in other data formats, such as shapefiles. For all feature classes, ArcMap updates the new feature's identifier attribute that keeps track of the number of features. Changes to other attributes, including user-defined IDs (like parcel numbers) or legal measurements of length and area, must be made manually.

Exercise 16b

The planning department has been notified of two pending property changes. In one case, an owner wants to subdivide her land so that she can sell a piece of it. In the other, the owner of one half of a duplex has bought the other half. Because the duplex had two owners, there was a parcel boundary dividing the house. Now the boundary needs to be removed.

1   In ArcMap, open **ex16b.mxd** from the **C:\ESRIPress\GTKArcGIS\Chapter16** folder.

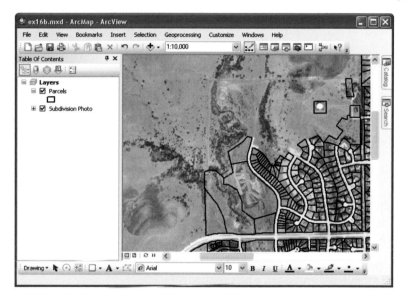

When the map opens, you see the familiar subdivision. You'll zoom in to the parcel to be split.

2   Click the Bookmarks menu and click First Parcel Site.

14
15
16
17

The display zooms in on parcel 2707. The yellow line (which appears in the book but not on your screen) shows where the parcel is to be split. When the parcel is split, ArcMap will update the attribute table with the correct area of the two new parcels. To see this for yourself, you'll check the area of parcel 2707 before you split it.

3 On the Tools toolbar, click the Identify tool and click parcel 2707. The orientation of your Identify window may look different from the graphic.

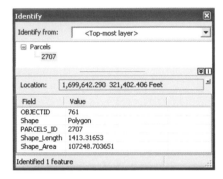

The Shape_Area field shows a value of 107,248 square feet before the decimal place.

4 Close the Identify window. In the table of contents, turn off the Subdivision Photo layer.

5 On the Standard toolbar, click the Editor Toolbar button.

6 On the Editor toolbar, click the Editor menu and click Start Editing. In the Create Features window, click Parcels, the only editable layer.

7 On the Editor toolbar, make sure the Edit tool is selected.

8 Double-click parcel 2707 to select it and display its edit sketch. You will draw the split line between the vertices indicated by red arrows in the graphic.

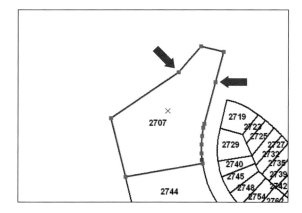

When you double-click to display a feature's edit sketch, certain tasks on the Editor toolbar are available to use such as the Reshape Features Tool and Cut Polygons Tool. In this exercise, you will use the Cut Polygons Tool.

Before digitizing the split line, you will set the snapping settings so that the line snaps to vertices.

9   On the Editor toolbar, click the Editor menu, point to Snapping, and click Snapping Tool. In the Snapping toolbar, click Vertex Snapping, as shown in the following graphic. Close the toolbar. (If you are continuing from the previous exercise, this may already be enabled.)

Next, you will choose the Cut Polygon Tool.

10   On the Editor toolbar, click the Cut Polygon Tool.

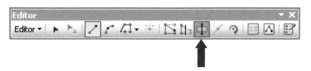

Although the parcel vertices no longer display when the task is changed, the split line you digitize will still snap to them.

11 On the display, the cursor changes to a crosshair. Move the cursor to the position shown in the following graphic. When it snaps to the vertex, click the left mouse button.

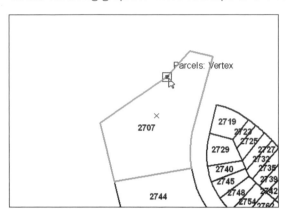

12 Move the cursor over the second vertex. When it snaps to it, double-click to end the split line. If you moved your cursor away from the vertex, press F2 on your keyboard to finish the sketch.

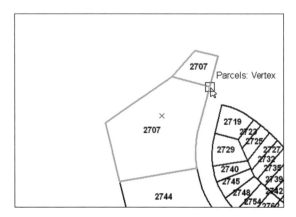

The parcel is split in two. Each parcel is labeled with the same PARCELS_ID of 2707. In the next exercise, you'll update the PARCELS_ID for the smaller parcel.

13 On the Editor toolbar, click the Editor menu and click Save Edits.

14 On the Tools toolbar, click the Identify tool. Click each of the two new parcels.

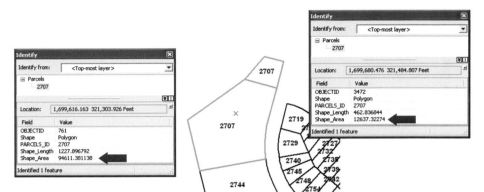

The sum of the areas of the two new parcels equals the area of the original parcel. The PARCELS_ID number, a user-defined attribute, does not change. The new parcel is assigned a new OBJECTID number.

Your next task is to merge two parcels where the owner of one half of a duplex has bought the other half.

15 Close the Identify window. Click the Bookmarks menu and click Second Parcel Site.

16 Turn on the Subdivision Photo layer.

In the center of the map display, two parcels—2855 and 2861—are located on a cul-de-sac. The parcel boundary crossing the duplex needs to be removed.

14
15
16
17

17 On the Editor toolbar, click the Edit tool.

18 Click anywhere on parcel 2855 to select it. Hold down the Shift key and click parcel 2861 to select it as well.

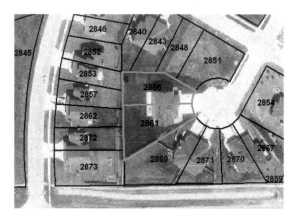

19 On the Editor toolbar, click the Editor menu and click Merge. In the Merge dialog box, click Parcels - 2855, as shown in the following graphic, then click OK.

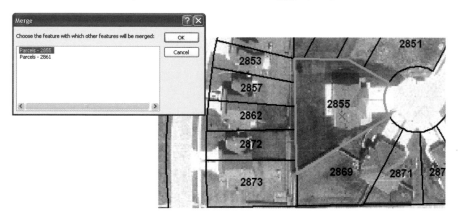

The two parcels become one. The merged parcel retains the attributes of parcel 2855.

20 In the table of contents, right-click the Parcels layer and click Open Attribute Table. Click the Show selected records button at the bottom of the table to display selected records.

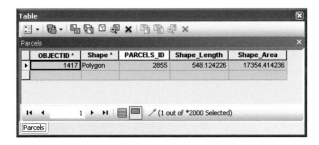

The record that remains is for the feature you chose in the Merge dialog box. Its Shape_Length and Shape_Area values are updated.

21 Close the table. On the Editor toolbar, click the Editor menu and click Stop Editing. Click Yes to save your edits. Close the Editor toolbar.

You don't need to save the map document because your edits were made directly to the Parcels feature class in the geodatabase.

22 If you are continuing to the next exercise, leave ArcMap open. Otherwise, exit the application. Click No when prompted to save your changes.

14
15
16
17

Editing feature attribute values

In ArcMap, you can change attribute values one at a time by typing new values into the table during an edit session. You can also change values for a selected set of records or for all records in a table with the Field Calculator. As you know from chapter 12, the Field Calculator assigns values to a field according to an expression, which may simply be a value, like "101" or "Residential," or may be an expression using math operators. The Field Calculator can be used either in or out of an edit session.

Besides editing attribute values, you can add and delete records or add new fields to a table and define properties for those fields. Records are added and deleted during an edit session. Fields are added and deleted outside an edit session.

Exercise 16c

Having split parcel 2707, you need to assign a new PARCELS_ID number to one of the new parcels. After that, you will add an attribute to the parcels feature class. In a previous meeting with the City Commission, you found that parcel sizes were often given as acres rather than square feet. The Parcels feature class doesn't have an acreage attribute. You will add one and have ArcMap calculate values for it by converting square feet to acres.

1   In ArcMap, open **ex16c.mxd** from the **C:\ESRIPress\GTKArcGIS\Chapter16** folder.

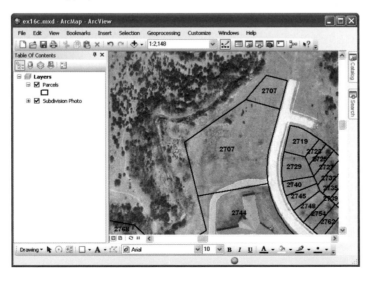

The map shows the parcels you split in the previous exercise. Both have an ID of 2707. You will assign a new ID number to the smaller parcel.

2　On the Standard toolbar, click the Editor Toolbar button.

3　On the Editor toolbar, click the Editor menu and click Start Editing.

The target layer is Parcels since it is the only layer available to edit.

4　On the Editor toolbar, click the Edit tool if necessary.

5　Click the small parcel 2707 to select it.

6　In the table of contents, right-click the Parcels layer and click Open Attribute Table. Click the Show selected records button at the bottom of the table.

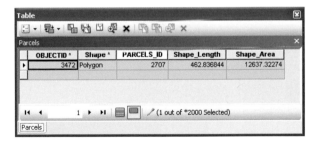

Field names on gray backgrounds indicate fields that are maintained by ArcMap. They cannot be edited. Names on white backgrounds indicate fields that can be edited.

14

15

16

17

7   Click in the cell for the PARCELS_ID field. Highlight and delete the existing value of 2707. Type **4001** and press Enter.

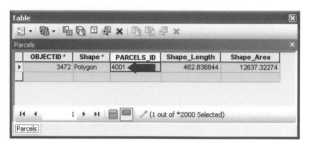

The value 4001 does not duplicate an existing parcel ID. This is the only value you need to edit. If there were other attributes, such as an address or owner name, you would update these as well.

8   On the Editor toolbar, click the Editor menu and click Stop Editing. Click Yes to save your edits. Move the table away from the map display.

When the edit session is finished, there are no longer any selected records in the table or selected features on the map. On the map, the smaller parcel is labeled with its new PARCEL_ID.

Now you will add a field to the table to store parcel size in acres.

9 At the bottom of the Parcels attributes table, click the Show all records button.

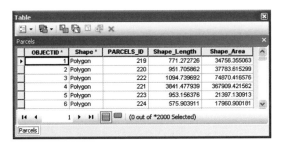

Currently, the Shape_Area field contains the square footage of each parcel. After adding the new field, you will write an expression in the Field Calculator to convert square feet to acres.

10 Click the Table Options menu and click Add Field. (If the Add Field command is disabled, you may not have concluded the edit session in step 8.)

In the Add Field dialog box, you name the new field, specify its data type, and set field properties. (This is equivalent to what you did in chapter 14 when you added fields to the WaterLines feature class in ArcCatalog.)

Adding fields from ArcCatalog

New fields can be added from ArcCatalog or ArcMap. With an ArcView license, fields can be added only to file and personal geodatabase or shapefile feature classes. With an ArcEditor or ArcInfo license, fields can also be added to feature classes in coverages and ArcSDE geodatabases. For more information, click the Contents tab in ArcGIS Desktop Help and navigate to *Professional Library > Data Management > Geographic data types > Tables > Creating and editing tables > Adding fields*.

11 In the Name text box of the Add Field dialog box, type **Acres**. Click the Type drop-down arrow and click Float, as shown in the following graphic. (Float is a numeric field that allows decimal points.) Click OK.

The new field is added to the table. Its values are set to <Null>.

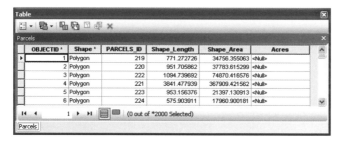

Now you will calculate values for the Acres field, which you can do whether you are in an edit session or not. Field calculations made in an edit session take longer but have an Undo option. Calculations made outside an edit session are faster but can't be undone. (To replace an incorrect calculation, you can make a new one.) For the sake of speed, you will make your calculation without starting an edit session.

12 In the attribute table, right-click the Acres field name and click Field Calculator.

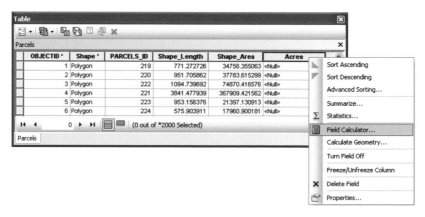

The Field Calculator opens.

You'll create an expression to convert feet to acres. Since one acre equals 43,560 square feet, dividing the values in the Shape_Area field by 43,560 will give each parcel's acreage.

13 In the Fields box of the Field Calculator, double-click Shape_Area to add it to the expression box. Click the division (/) button. Type a space and type **43560**. Make sure your dialog box matches the following graphic, then click OK.

The values are calculated and written to the Acres field of the table.

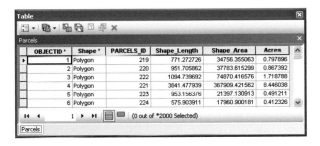

When you edit a feature in a geodatabase feature class, ArcMap automatically updates its Shape_Length and Shape_Area values. ArcMap does not, however, manage fields that you have added yourself. Therefore, when you make changes to parcels that affect their size, you'll have to recalculate the Acres field.

You don't need to save the map document. Your edits were made directly to the feature class attributes in the geodatabase.

14 Close the attribute table. Close the Editor toolbar. If you are continuing to the next chapter, leave ArcMap open. Otherwise, exit the application. Click No when prompted to save your changes.

Chapter 17

Geocoding addresses

Creating an address locator
Matching addresses
Rematching addresses

In chapter 15, you learned that much spatial data was originally digitized from paper maps. It's also possible to create spatial data—specifically, point features—from information that describes or names a location. The most common kind of information that does this is an address. The process of creating map features from addresses, place-names, or similar information is called geocoding.

The value of geocoding is that it lets you map locations from data that is readily available. If you own a business, you can't buy an atlas that has a map of your customers. With a list of your customer addresses, however, you can make this map for yourself.

Geocoding requires that you have an address table—a list of addresses stored as a database table or a text file. You also need a set of reference data, such as streets, on which the addresses can be located.

Address	City	State	Zip
3387 Arlington Avenue	Riverside	California	92506
5314 Victoria Avenue	Riverside	California	92506
2466 Fairview Avenue	Riverside	California	92506
2464 Arlington Avenue	Riverside	California	92506
2200 Arroyo Drive	Riverside	California	92506

Address table

Reference data

ArcGIS uses address information in the attribute table of the reference data to figure out where to locate address points. The more detailed the reference data, the more accurately addresses can be located.

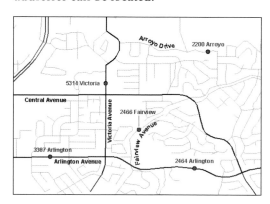

The same table of addresses is geocoded using two sets of reference data. On the street map of Riverside, California, address points are placed in the correct location on the correct street. On the ZIP Code map of the United States (zoomed in to Southern California), the same address points are placed in the center of the correct ZIP Code. Because all five points have the same location, they look like one point.

Street-level reference data is commercially available. Its attributes include street name, street type (Avenue, Boulevard, and so on), and directional prefixes and suffixes necessary to avoid ambiguity in address location. Each street feature is divided into segments that have beginning and ending addresses, much as you see on neighborhood street signs. This makes it possible to estimate the position of an address along the length of a street segment. There may be separate address ranges for each side of the street, so an address can be placed on the correct side of the street.

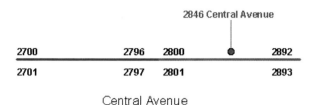

To geocode the address 2846 Central Avenue, ArcGIS examines the reference data and finds the street segment of Central Avenue that has the correct address range. Since 2846 is midway between 2800 and 2892, the address point is placed in the middle of the street segment, on the even-numbered side of the street.

In addition to an address table and reference data, geocoding requires an address locator. An address locator is a file that specifies the reference data and its relevant attributes, the relevant attributes from the address table, and various geocoding rules and tolerances.

The output of the geocoding process is either a shapefile or a geodatabase feature class of points. The geocoded data has all the attributes of the address table, some of the attributes of the reference data, and, optionally, some new attributes, such as the x,y coordinates of each point.

14

15

16

17

Creating an address locator

Geocoding starts with creating an address locator. Address locators come in different styles, each appropriate to reference data with different attributes. For example, the style called "Single Field" is used with reference data that contains only a single attribute of geographic information. In a reference map of the United States, for instance, this would probably be a state name field. The data to be geocoded could be located in the correct state, but no more precisely than that.

The address locator style called "ZIP" is used with reference data that has a ZIP Code attribute. A table of U.S. addresses could be located within the correct ZIP Codes.

The "US Streets" style is used with street reference data that contains a street name attribute and beginning and ending address ranges for each side of a street. A table of U.S. addresses could be geocoded to their approximate positions along the length of a street and on the correct side of the street.

Each address locator style requires that certain attributes be present in the address table as well. For more information about address locator styles, click the Contents tab in ArcGIS Desktop Help and navigate to *Professional Library* > *Data Management* > *Geographic data types* > *Locators* > *Working with address locators and geocoding*.

Exercise 17a

You work for an online business directory that sells advertising—everything from listings and banner ads to interactive storefronts. A new service you plan to offer is a digital map that displays the location of a client's business. Your job is not to create the Internet map service itself but rather to test the technology for converting customer addresses to map points. Being inexperienced with geocoding, you'll work with a small sample of your potential customer base.

1   Start ArcMap. In the ArcMap—Getting Started dialog box, under the Existing Maps section, click Browse for more. (If ArcMap is already running, click the File menu and click Open.) Navigate to **C:\ESRIPress\GTKArcGIS\Chapter17**. Click **ex17a.mxd** and click Open.

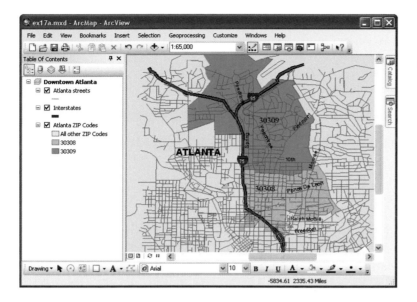

The map shows a portion of downtown Atlanta, Georgia, with layers of streets, interstate highways, and ZIP Codes.

2 At the top of the table of contents, click the List By Source button.

When List By Source is selected, the table of contents shows nonspatial tables, such as dBASE files, that belong to the map document. In this case, you see a table of customers, which contains a sample of your customer database. This is the address table you'll geocode in the next exercise.

The address locator style you choose depends on the attributes in your reference layer (the Atlanta streets layer) and in your address table. You'll open the two tables to look at this information.

3 In the table of contents, right-click the Customers table and click Open.

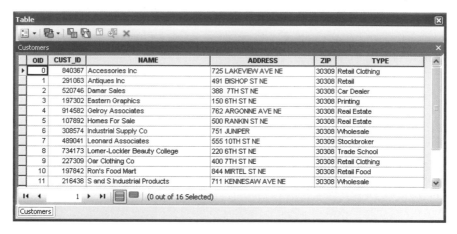

The attributes include customer ID, name, street address, five-digit ZIP Code, and business type. For geocoding, the relevant attributes are ADDRESS and ZIP.

4 Close the Attributes of Customers table. In the table of contents, right-click the Atlanta streets layer and click Open Attribute Table.

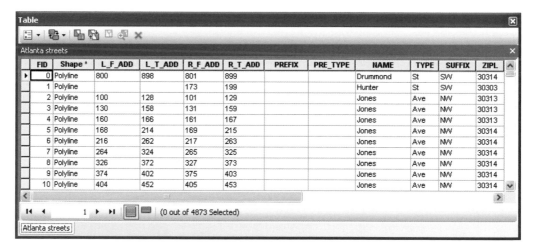

The four address range attributes are L_F_ADD (left "from" address), L_T_ADD (left "to" address), R_F_ADD (right "from" address), and R_T_ADD (right "to" address). These are the beginning and ending street numbers on each side of a street segment.

There are several other address attributes:

- PREFIX is a direction that precedes a street name, like the "N" in 139 N. Larchmont Blvd.
- PRE_TYPE is a street type that precedes a street name, like "Avenue" in 225 Avenue D.
- ZIPL and ZIPR are the ZIP Codes for the left and right sides of a street segment.
- CITYL and CITYR are the city names for the left and right sides of a street segment.
- NAME, TYPE, and SUFFIX direction are the familiar attributes of a street address.

The attributes in the Atlanta streets table most closely match the "US Address–Dual Ranges" address locator style.

5 Close the table. On the Standard toolbar, click the Catalog Window button. (Your Catalog window may be docked or floating.)

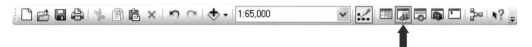

6 In the catalog tree, navigate to **\GTKArcGIS\Chapter17\MyData**. Right-click **My Data**, point to New, and click Address Locator.

14
15
16
17

The Create Address Locator window appears.

7   Click the Browse button next to the Address Locator Style box. In the Select Address Locator Style dialog box, click US Address–Dual Ranges, and click OK.

The Dual Ranges style is used for address ranges that have data for both sides of a street segment. Below, the Field Map section populates with fields that are required using the Dual Ranges locator style. Now you'll specify the reference data that will be used with it.

8 Click the Reference Data drop-down arrow and click Atlanta streets. Make sure your dialog box matches the following graphic.

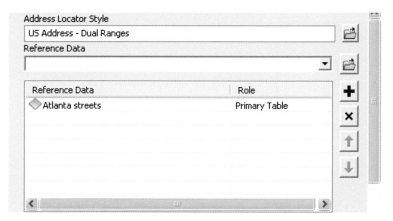

The reference data is added to the list and has a Primary table role.

9 Scroll down to view the Field Map list.

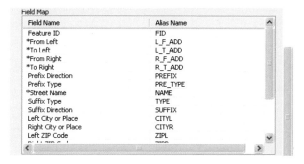

The Field Map list shows the types of address information in the reference data that the address locator style can use. Required information is represented with an asterisk in front of the field name. The values in the corresponding drop-down lists show the reference data attributes that contain this information. In each case, ArcGIS has already selected the correct field name. If it hadn't, you could use the drop-down arrows to choose the corresponding fields.

14
15
16
17

10 Click the Browse button next to the Output Address Locator box. In the dialog box, navigate to **\GTKArcGIS\Chapter17\MyData** and type **Atlanta customers** into the Name box. Make sure your dialog box matches the following graphic, then click Save.

11 Make sure your dialog box matches the following graphic, then click OK.

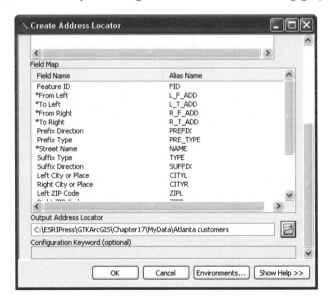

When the Create Address Locator tool completes, a pop-up notification appears by the system task bar. Click the link if you want to view the progress report.

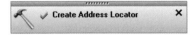

ArcGIS creates the new address locator in the output folder specified.

In the next exercise, you'll add the Atlanta customers address locator to ArcMap and use it to match addresses.

12 If you're continuing with the next exercise, leave ArcMap open. Otherwise, exit the application. Click No when prompted to save your changes.

14
15
16
17

Matching addresses

Now that you've created an address locator, you can compare the table of addresses to the reference data attributes and look for matches.

The first step in address matching is standardizing the addresses in the address table; that is, dividing them into their component parts. This is done by the address locator. For example, the address "500 Rankin Street NE" has four parts: the street number (500), the street name (Rankin), the street type (Street), and a suffix (NE). The reference data is already standardized—each address part is a separate attribute.

ArcGIS takes each standardized address from the address table and looks for features in the reference data with matching parts. For each address, it generates a list of probable or possible matching locations, called candidates, in the reference data. Each candidate gets a score that rates its likelihood of being the correct location. For ArcGIS to make a match and create a point feature, a candidate's score must reach a certain level. You can choose both the score that defines a location as a candidate and the score that defines a candidate as a match. By default, the minimum candidate score is 10, on a scale of 100, and the minimum match score is 60.

When ArcGIS has made as many matches as it can, it creates a new dataset, which has a point feature for each address, whether matched or not. Features that have unmatched addresses are not assigned spatial coordinates, and therefore do not display. You can adjust the geocoding options and try to find matches for unmatched addresses.

Most of the time, geocoding is done from a table of addresses, but as long as you have created an address locator, you can also geocode by typing address information into the Find dialog box. Geocoding this way does not create new features, but will flash a matching location on the map or place a graphic there. It's useful in situations where you need to locate a single address or a few addresses quickly—for example, to make immediate deliveries.

Exercise 17b

Before geocoding your customer table, you'll test the address locator by locating the address of the Ace Market, a downtown Atlanta business.

1 In ArcMap, open **ex17b.mxd** from the **C:\ESRIPress\GTKArcGIS\Chapter17** folder.

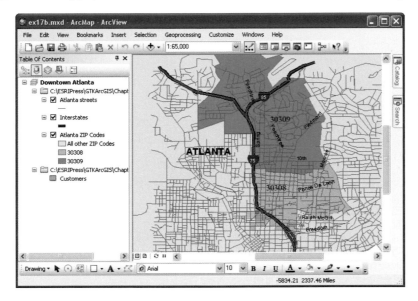

The map shows the layers of Atlanta streets, interstate highways, and ZIP Codes. The Customers table is at the bottom of the table of contents. (If you don't see it, click the List By Source button at the top of the table of contents.)

2 Click the Customize menu, point to Toolbars, and click Geocoding.

The Geocoding toolbar appears.

You'll add the address locator to the map document.

3   On the Geocoding toolbar, click the Address Locators drop-down arrow and select <Manage Address Locators…>.

The Address Locator Manager window appears.

4   Click Add. In the Add Address Locator browser, navigate to **\GTKArcGIS\Chapter17 \MyData** folder. Click the **Atlanta customers** address locator you created in the previous exercise. Make sure your dialog box matches the following graphic, then click Add.

The Atlanta customers address locator and a description of the style appear in the Address Locator Manager window.

5 Make sure your window matches the following graphic, then click Close.

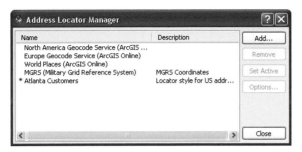

The asterisk beside Atlanta customers signifies that it is the active address locator. Now that the address locator has been added to the map document, you can find addresses on the map.

6 On the Tools toolbar, click the Find button.

7 In the Find dialog box, click the Locations tab. The Atlanta customers address locator should be displayed in the Choose an address locator drop-down list. If it isn't, click the drop-down arrow and select it.

8 In the Full Address box, type **1171 Piedmont Ave**, the address of the Ace Market. Make sure your dialog box matches the following graphic, then click Find.

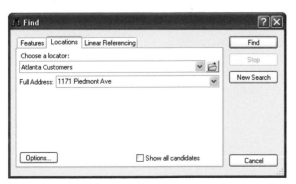

14

15

16

17

One candidate street segment appears at the bottom of the dialog box. Its score is 94.

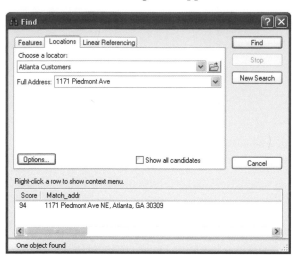

By default, only the candidate with the highest score is displayed. If you wanted, you could display additional candidates by checking the Show all candidates check box.

9   Scroll across the dialog box and look at the candidate's address information.

The address range for odd street numbers, RightFrom and RightTo, is correct because it includes the street number 1171. The street name, street type, and suffix directions are also correct.

A score of 94 is high, but to be sure, include more information for the Ace Market address: the suffix and ZIP Code.

10   In the Find dialog box, click the Full Address box and type **1171 Piedmont Ave NE 30309**. Make sure your dialog box matches the following graphic, then click Find.

The candidate, listed first, now gets a perfect score. You could display all candidates by checking the Show all candidates check box.

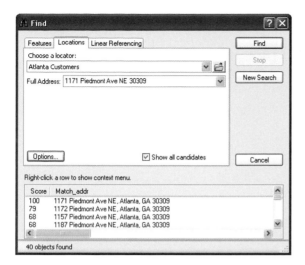

11 Click the first candidate to highlight it. Right-click anywhere on the highlighted candidate. On the context menu, click Add Point. Close the Find dialog box.

ArcGIS places a black dot on the map at the location of the candidate.

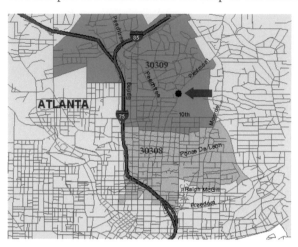

Since you geocoded the Ace Market only to make sure that the address locator was working, you can now delete the graphic.

12 On the Tools toolbar, click the Select Elements tool.

13 Click the graphic to select it. (Blue selection handles appear.) Press the Delete key on your keyboard.

Now you'll geocode the customer table.

14 On the Geocoding toolbar, click Geocode Addresses.

In the Choose an Address Locator to use dialog box, the Atlanta customers address locator is highlighted.

15 Click OK to open the Geocode Addresses dialog box.

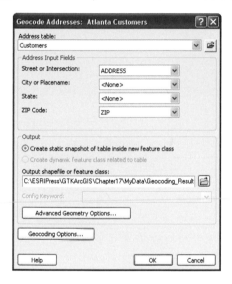

The City or Placename and State drop-down lists are specified as <None> because the Customers table does not contain that information.

Geocoding creates a point feature for every address. The points can be saved as a shapefile or as a geodatabase feature class. Your Atlanta data is in shapefile format, so you'll save the address points as a shapefile, too.

16 Click the Browse button next to the Output shapefile or feature class box.

14

15

16

17

Exercise **17b** Matching addresses

17 In the Saving Data dialog box, navigate to **\GTKArcGIS\Chapter17\MyData**. Make sure the Save as type drop-down list is set to Shapefile. In the Name box, highlight the default name of Geocoding_Result.shp and type **Geocode1.shp**. Make sure that your dialog box matches the following graphic, then click Save.

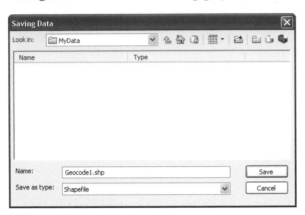

The Geocode Addresses dialog box updates with the new path.

18 Make sure your dialog box matches the following graphic, then click OK.

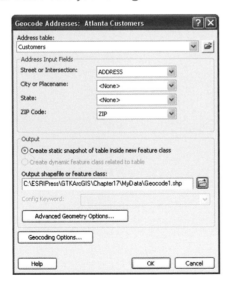

ArcMap geocodes the addresses and creates the output shapefile. When the process is done, a results window appears.

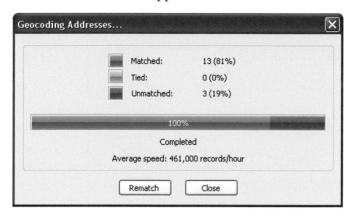

Of the sixteen addresses in the Customers table, thirteen were matched. Three addresses were unmatched. You'll match the unmatched addresses in the next exercise. For now, you'll look at the attributes of the geocoded shapefile.

19 In the Geocoding Addresses window, click Close.

The Geocoding Result: Geocode1 layer is listed at the top of the table of contents. There are thirteen point features displayed on the map, representing the locations of the matched addresses.

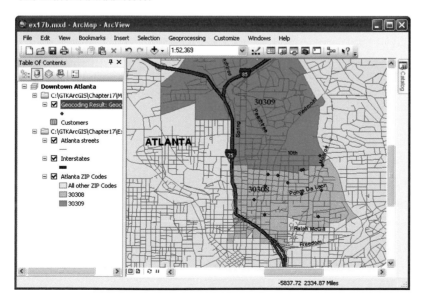

20 In the table of contents, right-click the Geocoding Result: Geocode1 layer and click Open Attribute Table.

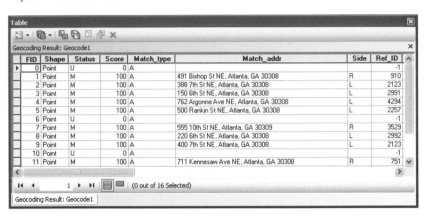

The table contains sixteen records, one for each address in the Customers table. The Status attribute tells you whether or not each address was matched (M), unmatched (U), or tied (T) with another address for the highest score. Point geometry is created only for records with the status M. The Score attribute contains the match score and the Side attribute tells you whether an address is on the right or left side of the street.

The next two attributes, ARC_Street and ARC_ZIP, contain the information from the ADDRESS and ZIP fields in the Customers table. The remaining attributes are the same as those in the Customers table.

Although the ARC_Street and ARC_ZIP attributes contain the same values as ADDRESS and ZIP, they serve a different purpose. When you rematch unmatched addresses, you can edit the values in ARC_Street and ARC_ZIP to make matches. You would do this, for example, if you found mistakes in the Customers table. The ADDRESS and ZIP attributes, on the other hand, preserve the original information from the Customers table.

21 Close the table. You have already saved the geocoded shapefile. There is no need to save the map document.

22 If you're continuing with the next exercise, leave ArcMap open. Otherwise, exit the application. Click No when prompted to save your changes.

Rematching addresses

When you geocode at the street address level, it's common to have addresses that aren't matched. Occasionally, this is the result of reference data that has errors or is incomplete. For instance, an address may belong to a subdivision that is newer than the reference data. More often, there is a mistake, such as a spelling error, in the address table. In some cases, the problem is with the way ArcGIS has standardized an ambiguous address.

Unmatched addresses can be rematched automatically (ArcGIS does all of them at once) or interactively (you do them one at a time). If you rematch automatically, you need to adjust the matching options, particularly the minimum match score and the spelling sensitivity, or your results won't change. If you rematch interactively, you have additional options. You can edit the way ArcGIS has standardized an address, and you can match a candidate with a score below the minimum match score.

Exercise 17c

So far, the address locator has found matches for thirteen addresses in the Customers table, leaving three unmatched. In this exercise, you'll match these three. First, you'll modify the address locator's settings to generate more candidates. Then you'll interactively match each address.

1 In ArcMap, open **ex17c.mxd** from the **C:\ESRIPress\GTKArcGIS\Chapter17** folder.

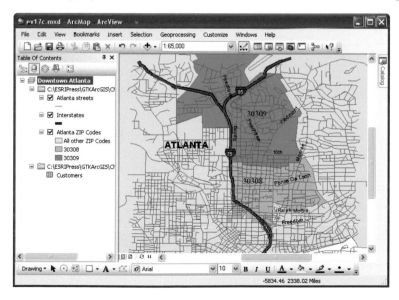

14
15
16
17

The map layers include the familiar streets, interstate highways, and ZIP Codes of downtown Atlanta. If necessary, click the List By Source button at the top of the table of contents.

Now you'll add the layer of geocoded addresses you created in the previous exercise.

2 On the Standard toolbar, click the Add Data button.

3 In the Add Data dialog box, navigate to **\GTKArcGIS\Chapter17\MyData**. Click **Geocode1.shp** as shown in the following graphic, then click Add.

ArcMap warns you that the layer is missing spatial reference information.

4 Click OK on the warning message.

The Geocode1 layer is added to the table of contents. It is symbolized with the default ArcMap point symbol in a random color. If you like, you can change the symbol to make it easier to see. In the following graphics, the point symbol has been set to Circle2, Fire Red, 7 points.

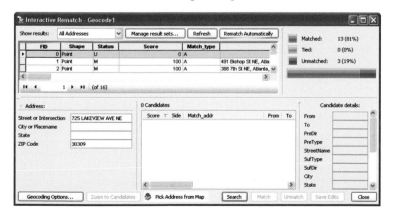

Now you'll rematch the unmatched addresses.

5   In the Table of contents, click the Geocode1 layer. On the Geocoding toolbar, click
    Review/Rematch Addresses. (If the Geocoding toolbar is not visible, click the
    Customize menu, point to Toolbars, and click Geocoding.)

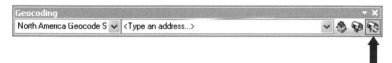

The Interactive Rematch dialog box opens.

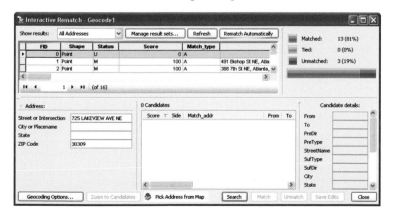

In the Statistics panel in the upper right corner, you see the results from the last exercise.
The Show results option at the top left allows you to specify which addresses you want
to rematch.

6 Click the Show results drop-down arrow and choose Unmatched Addresses.

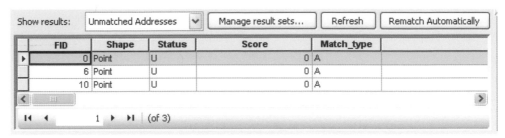

The table displays the three unmatched records with scores of 0. The first one is high-lighted in blue. Below the table, the Street or Intersection box and the ZIP Code box contain the current address information for the highlighted record. Edits to this information will be written to the ARC_Street and ARC_ZIP fields of the Geocode1 table. You will not use the City or Placename and State boxes.

The Standardized Address field shows how ArcGIS has divided the address into parts. The empty white box displays match candidates for the highlighted record. There are no match candidates for the first record.

In the address table, scroll to the right and notice that the ZIP Code for the first record, 30309 is different from the ZIP Codes for the other records. It is possible that the wrong ZIP Code was entered in the Customers table. You will edit the ZIP Code to see if this produces any candidates.

7 In the ZIP Code box in the middle of the dialog box, highlight the value 30309 and type **30308**. Press Enter.

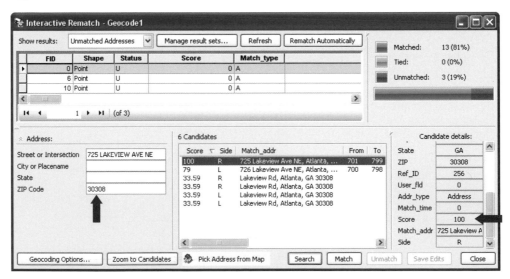

Six match candidates appear at the bottom of the dialog box. The top candidate has a perfect score of 100.

8 Make sure the first candidate is highlighted. Click Match.

The candidate is matched. At the top of the dialog box, the record's Status value changes from U to M and its ARC_ZIP value changes from 30309 to 30308. Fourteen addresses are now reported as matched.

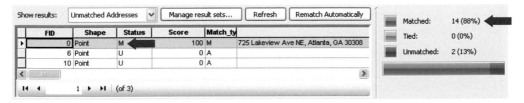

Now you'll match the other two addresses.

9 At the top of the Interactive Rematch dialog box, click the tab to the left of the second record to select it.

There are fourteen match candidates. Notice that the top match score is 84.73. This is the correct candidate to match the address to.

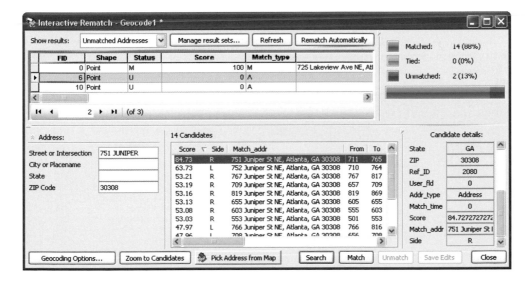

10 Make sure the first candidate is highlighted. Click Match.

The candidate is matched. Fifteen addresses are now reported as matched.

11 At the top of the Interactive Rematch dialog box, click the tab to the left of the third record to select it.

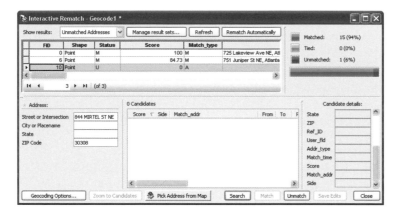

There are no match candidates. Notice the spelling of the street name, MIRTEL.

12 In the Street or Intersection box, highlight MIRTEL and type **MYRTLE**. Press Enter.

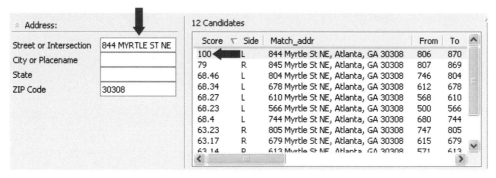

The scores of twelve candidates are displayed. The top candidate now has a score of 100.

13 Make sure that the top candidate is highlighted, then click Match.

The record's status changes from U to M, and its name is changed in the ARC_Street field.

You've matched all the addresses in your customer table.

The Statistics panel shows you that sixteen addresses (100%) were matched.

14 Close the Interactive Rematch dialog box.

On the map there are now sixteen geocoded points.

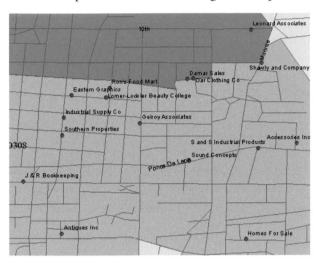

15 In the table of contents, right-click the Geocode1 layer and click Zoom to Layer. Right-click the layer again and click Label Features.

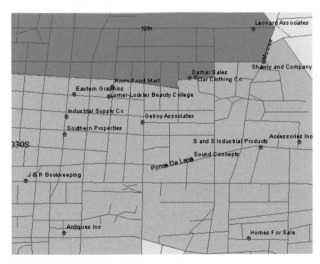

The points are labeled with the names of your customers' businesses.

16 In the table of contents, right-click the Geocode1 layer and click Open Attribute Table.

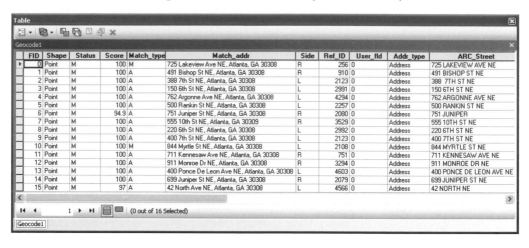

All the records have the value M in the Status field. Edits that you made during rematching ("Mirtel" to "Myrtle" for example) are reflected in the ARC_Street and ARC_ZIP fields.

Now that you know how to geocode addresses, the project to offer digital maps as one of your company's advertising services can move forward.

17 Close the table.

18 If you want to save your work, save it as **my_ex17c.mxd** in the **\GTKArcGIS\Chapter17 \MyData** folder.

19 If you are continuing to the next chapter, leave ArcMap open. Otherwise, exit the application. Click No when prompted to save your changes.

Chapter 18

Making maps from templates

Opening a map template
Adding x,y data to a map
Drawing graphics on a map

Making a map that's accurate, informative, and nice to look at usually takes time, as you'll see in the next chapter. Sometimes, however, you have to make a professional-quality map on short notice. ArcMap comes with a number of templates to help you do this. In a template, the map elements you need (data frames, legend, title, north arrow, background color, and so on) are already in place. All you do is add data and the map is ready to print.

You may not even need to add data—many ArcMap templates already contain layers for the world and the United States. A single template may be all you need, or you may want to use a template as a basemap and add your own layers to it.

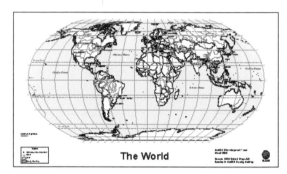

The LandscapeClassic template (left) has layout elements but no data. The WorldRobinson template (right) has countries, cities, rivers, and lakes in a Robinson projection.

Opening a map template

By default, every ArcMap document uses a template called the normal template. When you open a new document and switch to layout view, the layout page (also called the virtual page) is blank and has a single data frame. This is the appearance prescribed by the normal template.

In the course of making a map, you may add data to this data frame, resize it, insert more data frames, and add such elements as legends and scale bars. When you save the map, you save it not as a template but as a map document, a file with an .mxd extension. That way, the template does not change. Then, when you create a new ArcMap document, you start again with a blank page.

Rather than designing your own layout, you can open a custom template (that is, a template other than the normal template). Again, any changes you make are saved as a map document, so the template is not affected.

Template	A	B	C

 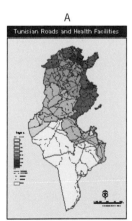  

On the left is the PortraitClassic template. On the right are three maps that use it, each with different data. In map A, a title has been specified but the template is otherwise unchanged. In map B, the legend, north arrow, and scale bar have been moved from their default positions. In map C, the background color has been changed.

18

19

You can also create new templates. If you've made an arrangement of map elements and data frames (with or without data) that you want to use again, simply save the map document (.mxd). Essentially any map document can be used as a template as long as it is opened as a template. ArcMap custom templates are stored in the \Bin\Templates folder of your ArcGIS installation path (for example, C:\Program Files\ESRI\Desktop10.0\ArcMap\Templates). You can add your own templates to this folder or to any folder you choose. New categories for templates are made available by creating folders beneath the templates folder locations. Categories will only appear in the Getting Started, New Map, and Select Template dialog boxes if a map document is present in the folder under the Templates or My Templates locations.

Beginning with ArcGIS 10, ESRI no longer creates map templates (.mxt) files (but can still open them.) For more information, click the Contents tab in ArcGIS Desktop Help and navigate to *Professional Library > Mapping and Visualization > What's new for map templates.*

Exercise 18a

It's early June 1999. As the information officer for the Philippine Atmospheric, Geophysical and Astronomical Services Administration, you've been asked to talk to a group of journalists about Typhoon Maggie, also known as Etang.

Typhoon Etang was first spotted off the coast of Samar, an island in the east central part of the Philippines. It has since traveled northwest, and is currently about 250 kilometers off the northeast coast of Luzon, where the capital, Manila, is located. The typhoon will probably pass harmlessly north of Luzon, through open ocean, but should it strike land it could be devastating.

The presentation is scheduled for tomorrow, but you weren't told about the briefing until today. Now it's 11:00 PM and you're trying to create a Microsoft PowerPoint slide presentation and still have time for a little sleep.

You want to show the path of the typhoon and its current location on a map of the country with rivers, populated places, and other useful geography. You'll use an ArcMap template for the basemap and draw the typhoon path yourself.

1   Start ArcMap. In the ArcMap–Getting Started dialog box, click Browse for more templates. (If ArcMap is already running, click the File menu and click New.)

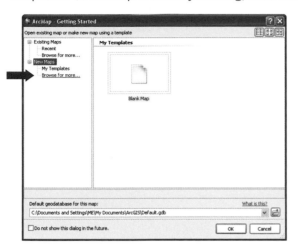

The Open dialog box appears. You'll use a template that was specially created for this exercise.

2   In the Open dialog box, navigate to **C:\ESRIPress\GTKArcGIS\Chapter18**. Click **ex18a.mxd** and click Open.

18

19

The ArcMap–Getting Started dialog box shows ex18a as a new map template.

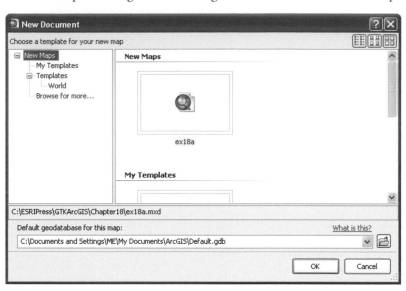

3   In the ArcMap–Getting Started dialog box, click OK.

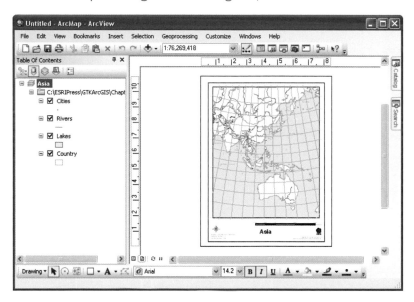

The map opens in layout view. You see the continents of Asia and Australia with cities, rivers, lakes, and country boundaries. A graticule marks lines of latitude and longitude.

Although the map is based on a template (.mxt file), you are now working in a new map document (.mxd file). Any changes you make are made to the map document, not to the template—this is what keeps the template reusable. (If you wanted to modify the template itself—which in this case you don't—you would click the File menu and click Open instead of New.)

With the entire map displayed, you can't read the text and labels. This is because the page in the window (the virtual page) is smaller than the printed page would be. The Layout toolbar shows you the size of the virtual page relative to the actual page. (Your percentage may be different.)

4   On the Layout toolbar, click the Zoom to 100% button.

The layout zooms to actual size and you see the map at the resolution it would have on a printed page. You can no longer see the entire layout.

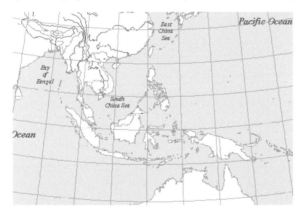

5   On the Layout toolbar (not the Tools toolbar), click the Pan tool.

18

19

6 Pan to the lower left corner of the map.

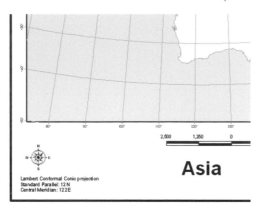

At 100 percent, map elements and text are legible, as they would be on the printed page.

7 On the Layout toolbar, click the Zoom Whole Page button.

The layout returns to full view.

The data frame shows all of Asia, but you are only interested in the Philippines. You'll use a bookmark to zoom in.

8 Click the Bookmarks menu and click Philippines.

The map zooms in on the Philippines. The graticule, with its relatively wide intervals of 10 degrees, will not help the journalists describe the typhoon's location. You'll reduce the size of the intervals to make the map more informative.

9 In the table of contents, double-click the Asia data frame to open the Data Frame Properties dialog box. Click the Grids tab.

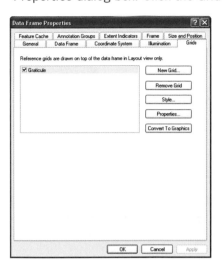

10 On the Grids tab, click Properties.

11 In the Reference System Properties dialog box, click the Intervals tab.

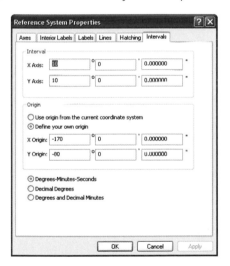

12 Replace the X Axis Interval value of 10 with **5** and the Y Axis Interval of 10 with **5**, as shown in the following graphic. Click OK, then click OK in the Data Frame Properties dialog box.

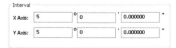

13 On the Tools toolbar, click the Select Elements tool.

14 Click outside the virtual page to unselect the data frame.

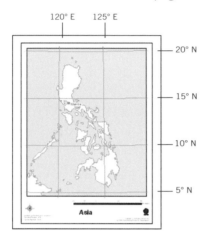

The graticule is now divided into intervals of 5 degrees. Parallels and meridians are labeled on the layout, but the labels are too small to see when you are zoomed to the whole virtual page.

To add a graticule to a data frame that doesn't have one, open the Data Frame Properties dialog box, click the Grids tab, and click New Grid to start the Grids and Graticules wizard. (Graticules mark latitude and longitude coordinates. Measured grids mark projected coordinates.) Grids and graticules display only in layout view.

You now have your basemap. In the next exercise, you'll display points that show Etang's progress.

15 If you want to save your work, click the File menu and click Save As. Navigate to **\GTKArcGIS\Chapter18\MyData**. Rename the file **my_ex18a.mxd** and click Save.

16 If you are continuing with the next exercise, leave ArcMap open. Otherwise, exit the application. Click No when prompted to save your changes.

Adding x,y data to a map

As you know from the previous chapter, ArcMap can turn geographic information from a table into points on a map. When the information consists of street addresses, the process is fairly elaborate. If, however, you have a table of coordinate values (obtained from a Global Positioning System device or other source), ArcMap can create point data with little preparation.

A table of coordinates must contain two fields, one for the x-coordinate and one for the y-coordinate. The values may be in any geographic or projected coordinate system.

Exercise 18b

As part of your presentation to the press, you want to show the location, speed, and path of Typhoon Etang. You have been given a text file of latitude–longitude coordinates showing where the typhoon has been at different times. You'll add this file to the map document and display the coordinates as points on the map.

1   In ArcMap, open **ex18b.mxd** from the **C:\ESRIPress\GTKArcGIS\Chapter18** folder.

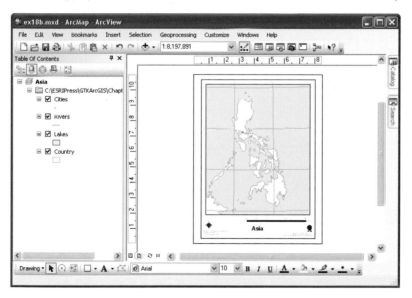

The map document opens in layout view and looks as it did at the end of the previous exercise.

You'll add the file of latitude–longitude coordinates to the map document.

18

19

Exercise **18b** Adding x,y data to a map

2 On the Standard toolbar, click the Add Data button.

3 In the Add Data dialog box, navigate to **\GTKArcGIS\Chapter18\Data**. Click **latlong_etang.txt**, as shown in the following graphic, and click Add.

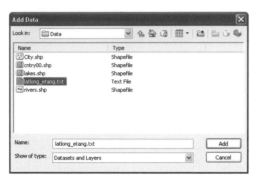

The text file is added to the table of contents and displayed in the List By Source view.

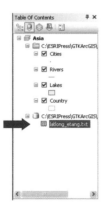

4 In the table of contents, right-click **latlong_etang.txt** and click Open.

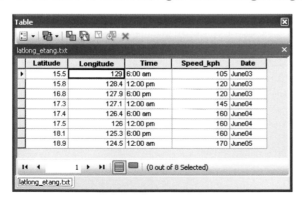

The table shows the typhoon's position and speed over the course of two days.

Although latitude and longitude are measured in degrees, minutes, and seconds, the values in the table have been converted to decimal degrees so that ArcMap can store and process them efficiently. Decimal degrees represent minutes and seconds as fractions. For example, the value 30°15' is 30.25 decimal degrees because 15 minutes is a quarter of a degree.

5   Close the table. In the table of contents, right-click latlong_etang.txt and click Display XY Data.

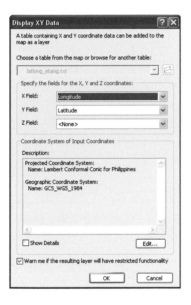

6   Click Edit under the Description window.

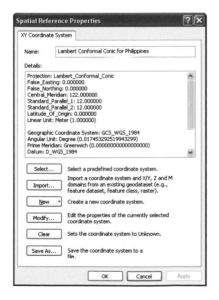

18

19

7   Click Import. In the Browse for Dataset dialog box, navigate to **\GTKArcGIS\Chapter18 \Data.** Click **City.shp**, then click Add. Click OK.

Note: Any of the shapefiles in this folder could have been used, since all have identical contents in their projection files.

8   Click the check box next to Show Details to see more information about the Coordinate System of Input Coordinates. Click OK.

9 Click OK in the Display XY Data dialog box to display the points on the map. Click OK to dismiss the "Table Does Not Have Object-ID Field" message.

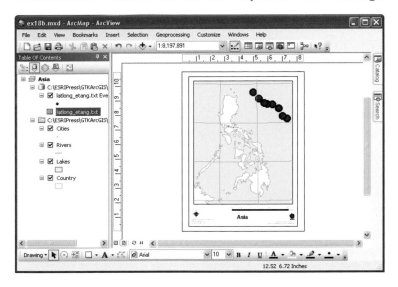

A layer called latlong_etang.txt Events is added to the table of contents. ("Events" is a technical term for points created from x,y coordinate values.)

Note: If the table on which an XY event layer is based does not have an ObjectID field, you get a warning message and you won't be able to perform certain tasks on the layer.

You'll give the layer a simpler name, then resymbolize the points with a symbol for typhoons.

10 In the table of contents, double-click the latlong_etang.txt Events layer. In the Layer Properties dialog box, click the General tab.

11 Replace the layer name with **Etang**, as shown in the following graphic, then click OK.

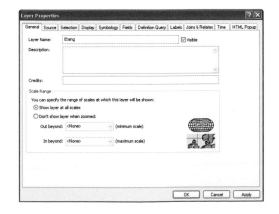

18
19

12 In the table of contents, click the symbol for the Etang layer. The Symbol Selector dialog box opens.

13 At the top of the Symbol Selector, type **Typhoon** into the search bar and press Enter. Under the Weather style heading, click the Typhoon symbol.

14 Change its size from 18 points to **4**. Make sure that your dialog box matches the following graphic, then click OK.

The points are resymbolized on the map.

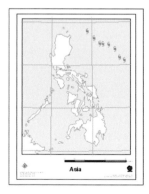

In the next exercise, you'll draw a line along the typhoon's path.

15 If you want to save your work, save it as **my_ex18b.mxd** in the **\GTKArcGIS\Chapter18 \MyData** folder.

16 If you are continuing with the next exercise, leave ArcMap open. Otherwise, exit the application. Click No when prompted to save your changes.

Drawing graphics on a map

With the tools on the Draw toolbar, you can add graphics and text to a layout. You should wait to do this until you're satisfied with the display scale because graphics, unlike features, do not scale proportionately as you zoom in or out on data. A graphic box that encloses a feature at one display scale, for instance, may not enclose it at another.

Exercise 18c

Now that you have added the point locations of Typhoon Etang to your map, you'll draw the typhoon's path by connecting the symbols. Then you'll add some descriptive text and change the map title. When your map is ready, you'll export it in a format that Microsoft PowerPoint supports so you can include it in your presentation.

1   In ArcMap, open **ex18c.mxd** from the **C:\ESRIPress\GTKArcGIS\Chapter18** folder.

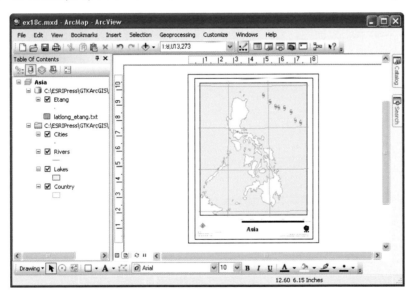

The map looks as it did at the end of the previous exercise. You'll zoom in on the layout to draw a line connecting the typhoon symbols.

18

19

2   On the Layout toolbar, click the Zoom In tool.

3   Drag a zoom rectangle corresponding to the red box in the following graphic.

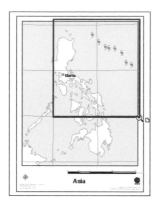

The layout zooms in to the area of the storm's path.

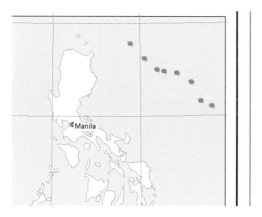

4   On the Draw toolbar, click the drop-down arrow by the Rectangle tool and click the Curve tool.

5 Draw the typhoon path on the map. Start by clicking the symbol farthest to the east. Click each symbol to draw a line connecting them. Double-click the last symbol to end the line. If you make a mistake, end the line, press the Delete key, and start again.

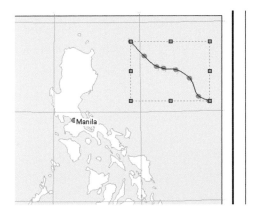

6 With the graphic selected, move the mouse pointer over the line and right-click. On the context menu, click Properties.

7 In the Properties dialog box, click the Symbol tab if necessary. Click the color square. On the color palette, click Electron Gold, as shown in the following graphic. Click OK in the dialog box.

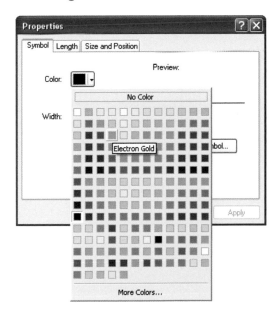

The new color is applied to the line.

18

19

8 Click outside the data frame to unselect the line.

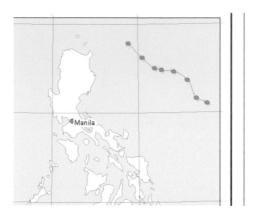

Now you'll add text to show the typhoon's time and speed at its last recorded position.

9 On the Tools toolbar, click the Identify tool.

10 Click the symbol at the last recorded position of Etang.

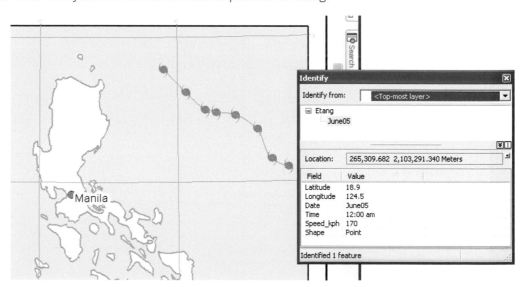

The time was midnight on June 5. The typhoon's speed was 170 kilometers per hour.

11 Close the Identify window. On the Draw toolbar, click the drop-down arrow by the New Text tool and click the Callout tool.

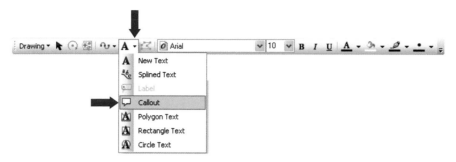

12 Click the same typhoon symbol again. In the pop-up text box, type **June 5 12:00 am - 170kph** and press the Enter key.

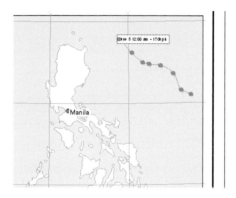

The text is added to the map and the text box is selected.

13 On the Draw toolbar, click the Font Size drop-down arrow and click 14.

18
19

14 Drag the callout text to the right of the typhoon symbol. Click outside the data frame to unselect the text.

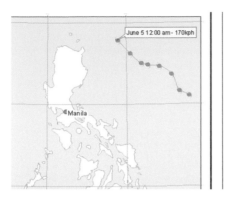

Since the map is no longer a map of Asia, you need to change its title.

15 On the Layout toolbar, click the Zoom Whole Page button.

16 Double-click the title Asia at the bottom of the map.

17 In the Properties dialog box, on the Text tab, replace Asia with **Path of Typhoon Etang**, as shown in the following graphic. Click OK.

18 Click outside the page to unselect the title.

Finally, you'll export an image of the map in EMF (.emf) format for your Microsoft PowerPoint slide presentation. EMF files can be added to common Windows applications such as Microsoft PowerPoint and Microsoft Word and resized without distortion. To learn more about the many graphic file formats ArcMap supports, click the Contents tab in ArcGIS Desktop Help and navigate to *Professional Library > Mapping and Visualization > Map export and printing > Exporting your map.*

19 Click the File menu and click Export Map.

20 In the Export Map dialog box, navigate to **\GTKArcGIS\Chapter18\MyData**. Click the Save as type drop-down arrow and click EMF (*.emf). Replace the default file name with **etang**, as shown in the following graphic, then click Save.

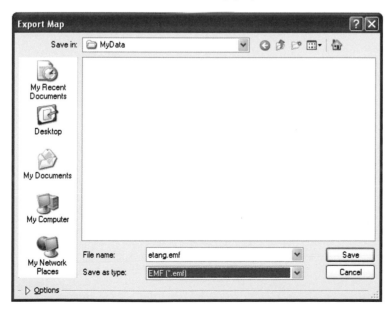

18

19

21 If you want to save your work, save it as **my_ex18c.mxd** in the **\GTKArcGIS\Chapter18 \MyData** folder.

22 If you are continuing to the next chapter, leave ArcMap open. Otherwise, exit the application. Click No when prompted to save your changes.

Chapter 19

Making maps for presentation

Laying out the page

Adding a title

Adding a north arrow, scale bar, and legend

Adding final touches and setting print options

A good map should inform, reveal, clarify, or convince. The elements for accomplishing these purposes include carefully prepared and symbolized data, a legend to explain the symbols, a descriptive title, projection information, and a source statement. A north arrow and scale bar often help to orient the map reader.

In this chapter, you'll create the map shown below of proposed tiger conservation areas in India.

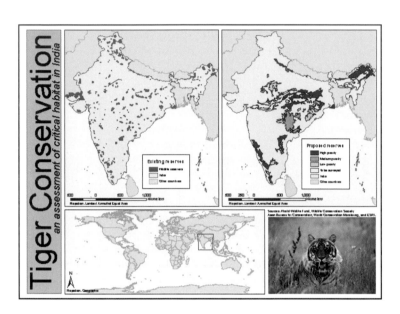

Laying out the page

The amount of space on which elements can be arranged may be anything from a letter-sized sheet of paper to a wall poster 44 by 34 inches. The orientation may be vertical (portrait) or horizontal (landscape). Choosing the dimensions and the orientation of the map before you start makes the layout process easier.

ArcMap has rulers, guides, and a grid to help you arrange map elements on a page. You can also align, nudge, distribute (space evenly), rotate, and resize selected elements to place them where you want.

Exercise 19a

You are a research associate for the World Wildlife Fund and have just finished an analysis of tiger populations in India. Loss of habitat is bringing wild tigers to the edge of extinction as forest lands are claimed for residential, commercial, and industrial use. Though poaching is not common in reserves or other protected areas, it is a problem in areas where there is little threat from the law.

Present wildlife reserves in India have been of some help, but are too small and scattered to maintain existing populations. After studying the data, you have developed a plan to connect existing reserves along vegetation corridors, consolidating them into larger areas that can better support the tigers.

The plan identifies the priority of proposed reserves as high, medium, or low. Reserves with higher priority have a better chance of sustaining a large tiger population for many years. The plan also identifies potential reserves for which research has yet to be completed.

You will create a map as part of a proposal to the United Nations Environment Program seeking support for the tiger conservation project.

1 Start ArcMap. In the ArcMap—Getting Started dialog box, under the Existing Maps section, click Browse for more. (If ArcMap is already running, click the File menu and click Open.) Navigate to **C:\ESRIPress\GTKArcGIS\Chapter19**. Click **ex19a.mxd** to highlight it and click Open.

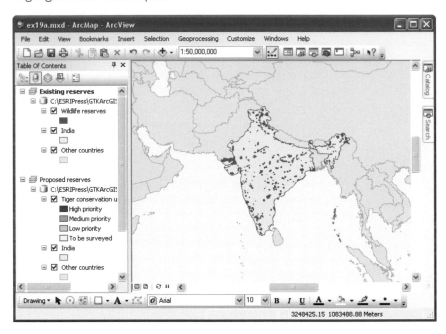

The map document contains three data frames: Existing reserves, Proposed reserves, and Overview. Existing reserves is active. Reserves are symbolized in green.

All three data frames will be part of your map layout. You'll show existing reserves next to proposed reserves. For orientation, you'll include a map of the world with India framed on it.

2 Click the View menu and click Layout View.

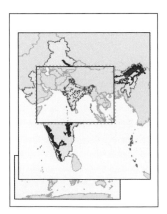

The data frames are stacked on the virtual page according to their order in the table of contents. It will be easier to see what you're doing if you make the ArcMap application window as large as possible.

3 In the ArcMap application title bar, click the middle button in the upper right corner to maximize the application window.

4 On the Layout toolbar, click the Zoom Whole Page button.

The virtual page is enlarged within the layout window. At the moment, its orientation is portrait.

5 Click the File menu and click Page and Print Setup.

6 In the Map Page Size frame of the Page and Print Setup dialog box, set the orientation to Landscape, as shown in the following graphic. Click OK.

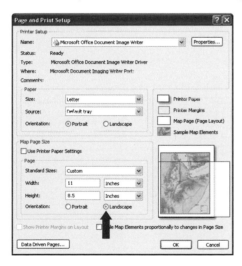

In the layout, the orientation changes. (You will also need to change the orientation of the printer paper, but you will do this later in the chapter.)

To help you place data frames exactly, you'll turn on guides. Guides are cyan lines originating from arrows on the layout rulers. They help you align and position elements but do not themselves appear on the printed map. In this exercise, some guides have already been set for you.

18

19

7 Click the View menu and click Guides.

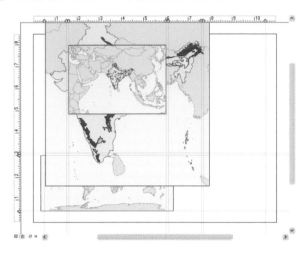

The guides display on the layout.

8 Place the mouse pointer over the vertical ruler on the left side of the display. Click to add a guide at 8 inches as shown in the following graphic.

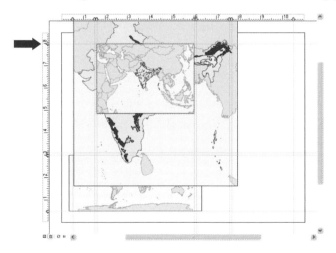

The guides will be more useful if map elements snap (automatically align) to them.

9 Right-click an empty part of the virtual page. On the context menu, point to Guides and click Snap to Guides.

Now you'll resize and arrange the data frames on the page, beginning with Proposed reserves (the largest data frame).

10 In the display area, click the Proposed reserves data frame to select it.

Once selected, it is outlined with a dashed blue line and marked with blue selection handles.

11 Move the mouse pointer over the data frame. The cursor has a four-headed arrow. Drag the frame so that its upper right corner snaps to the guides at 8 inches on the vertical ruler and 10.5 inches on the horizontal (top) ruler.

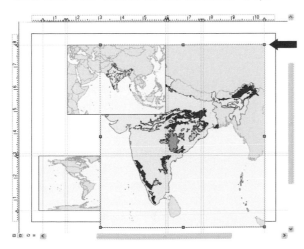

You'll make the data frame smaller by dragging a selection handle.

When you resize a data frame, ArcMap automatically adjusts the map extent and display scale. (If you want to make sure one or the other stays the same, you can set a fixed scale or extent on the Data Frame tab of the Data Frame Properties dialog box.)

12 Place the mouse pointer over the selection handle at the lower left corner of the data frame. The cursor changes to a two-headed arrow. Drag the corner of the data frame until it snaps to the guides at 3.1 inches (vertical ruler) and 6.1 inches (horizontal ruler).

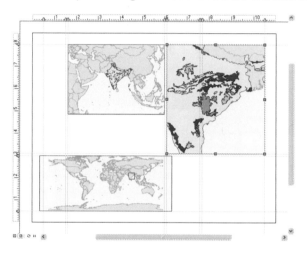

13 On the Standard toolbar, click the Scale box and type **28000000**.

Later, you'll use this number to set the scale of the Existing reserves data frame.

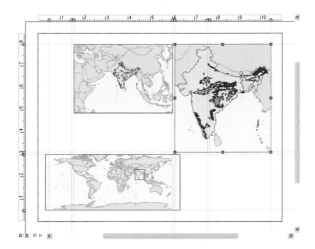

Next, you'll move the Overview data frame, which shows the map of the world with India marked by a red rectangle.

14 In the display area, click the Overview data frame to select it. Drag the frame so that its upper left corner snaps to the guides at 3 inches (vertical ruler) and 1.6 inches (horizontal ruler).

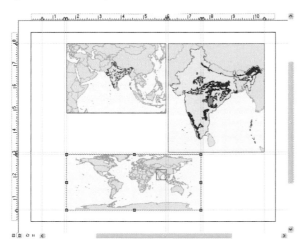

Now you'll resize the Existing reserves data frame to match the Proposed reserves data frame.

18

19

15 In the display area, click the Existing reserves data frame to select it. Put the mouse pointer over the selection handle at the bottom center of the data frame. The cursor changes to a two-headed arrow. Drag the data frame until its bottom edge snaps to the guide at 3.1 inches (vertical ruler).

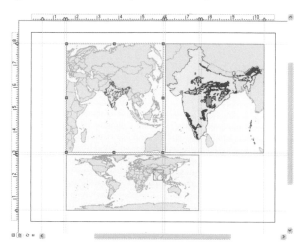

The data frames now have their final sizes and positions. (Another way to resize a data frame is with the Size and Position tab on the Data Frame Properties dialog box. The tab lets you set a height and width for the data frame and move a corner, midpoint, or the center of the frame to a page position you specify.)

In the Existing reserves data frame, India is displayed at a smaller scale than in the Proposed reserves frame. You'll change its display scale to match the Proposed reserves frame.

16 Make sure that the Existing reserves data frame is still selected. On the Standard toolbar, highlight the current view scale and replace it with the scale value of the Proposed reserves data frame. (If you didn't make a note of the value in step 12, type **28000000**.) Press Enter.

The display scale changes appropriately (yours may be slightly different). India, however, is not centered in the data frame.

17 On the Tools toolbar, click the Pan tool.

18 Pan the data so that India is similarly positioned in both data frames.

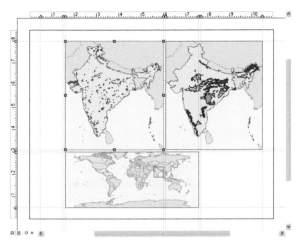

19 On the Tools toolbar, click the Select Elements tool.

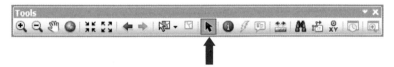

20 Click an empty part of the virtual page to unselect the data frame.

In the next exercise, you'll add a title to the map.

21 If you want to save your work, click the File menu and click Save As. Navigate to **\GTKArcGIS\Chapter19\MyData**. Rename the file **my_ex19a.mxd** and click Save.

22 In the ArcMap application title bar, click the middle button in the upper right corner to restore the application window to its former size.

23 If you are continuing with the next exercise, leave ArcMap open. Otherwise, exit the application. Click No if prompted to save your changes.

18

19

Adding a title

All maps have titles, and many have subtitles as well. A good title helps the reader understand what to look for in the map.

Exercise 19b

Now that you have laid out the three data frames, you'll add a title to convey the subject of the map (tiger conservation) and a subtitle to convey the specific focus (suitable habitat).

1   In ArcMap, open **ex19b.mxd** from the **C:\ESRIPress\GTKArcGIS\Chapter19** folder.

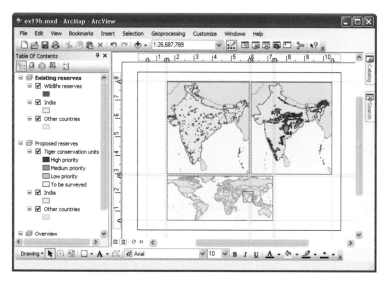

The map opens in layout view. It looks as it did at the end of the previous exercise.

2   In the ArcMap application title bar, click the middle button in the upper right corner to maximize the application window.

3   On the Layout toolbar, click the Zoom Whole Page button.

Now you'll insert the title.

4   Click the Insert menu, and click Title.

A text box is added to the page. The default title is the name of your ArcMap document, ex19b.mxd.

5   Double-click the text box. In the Properties dialog box, type **Tiger Conservation** and click OK.

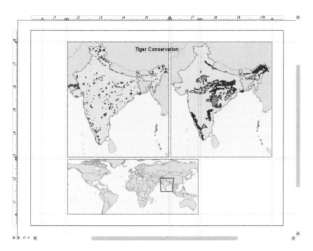

If you make a mistake and need to change the text, double-click the title again to open its text properties.

The title is too small. You can change its font, size, style, or color on the Draw toolbar.

6   On the Draw toolbar, for Font size, highlight the current value and replace it with **62**. Press Enter.

The title size changes to 62 points. Before positioning the title, you'll add a subtitle.

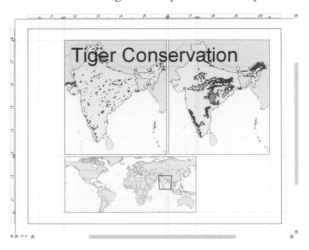

7   Click the Insert menu and click Title.

A text box is added to the page. Again, the default title is the map document name.

8   Double-click the text box. In the Properties dialog box, type **an assessment of critical habitat in India** and click OK.

You'll change the font and style of the subtitle.

9   On the Draw toolbar, click the Font Size drop-down arrow and click 22. Click the Italic button.

The subtitle is changed on the map.

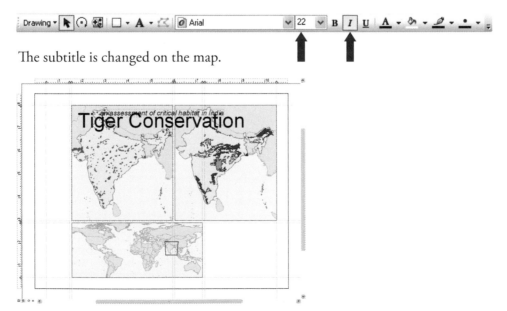

You'll position the two elements, then group them.

10 Drag the subtitle directly below the title, close but not touching. Position it so that the first word in the subtitle, "an," is underneath the "r" in "Tiger." Use the following graphic as a guide.

Tiger Conservation
an assessment of critical habitat in India

The subtitle needs to be very close to the title to fit in the space prepared for it.

11 Right-click the subtitle. On the context menu, point to Nudge and click Nudge Up. (You can also press 8 on your keyboard's numeric keypad.) Keep nudging until the subtitle is as close as possible to the title without touching it.

If you go too far, you can use Nudge Down (or press 2 on your numeric keypad). You can also delete the subtitle and try again.

12 With the subtitle selected, hold down the Shift key and click the title. Both elements are selected. Right-click either title. On the context menu, click Group.

Tiger Conservation
an assessment of critical habitat in India

Anything you do to the grouped title, such as changing the font size or color, will affect both elements. (To ungroup an element, select it, right-click it, and click Ungroup on the context menu.)

Now you'll rotate the grouped title and move it to the left side of the page.

13 With the grouped title selected, right-click it. On the context menu, point to Rotate or Flip and click Rotate Left.

It will be easier to put the title in the right place if it snaps to the guides.

14 Right-click an empty part of the virtual page. On the context menu, point to Guides and click Snap to Guides.

15 Drag the title to the left side of the virtual page so that the upper right corner snaps to the guides at 8 inches (vertical ruler) and 1.5 inches (horizontal ruler).

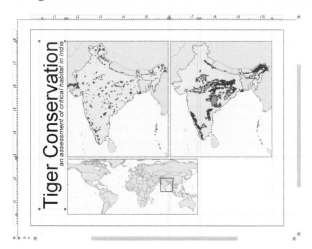

16 Click an empty part of the virtual page to unselect the grouped title.

Your title is not quite complete. You'll add a background rectangle to frame it.

17 On the Draw toolbar, click the Rectangle tool.

18 Drag a rectangle that snaps to the guides around the title. The upper left corner should snap to 8 inches (vertical ruler) and 0.5 inches (horizontal ruler). The lower right corner should snap to 0.5 inches (vertical ruler) and 1.5 inches (horizontal ruler).

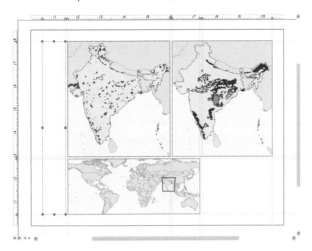

By default, the rectangle is pale yellow. You'll change its color and move it behind the title.

19 Make sure the rectangle is selected. On the Draw toolbar, click the Fill Color drop-down arrow. On the color palette, click More Colors.

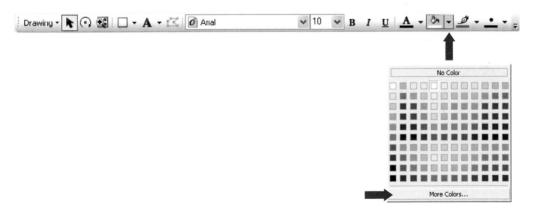

18
19

20 On the Color Selector dialog box, click the Color tab if necessary. Replace the R (red) value with **176**. Press the Tab key on your keyboard. Replace the G (green) value with **204** and press Tab again. Replace the B (blue) value with **204**. The custom color is previewed in the lower left corner of the dialog box. Make sure your dialog box matches the following graphic, then click OK.

The rectangle color changes to a blue gray.

21 Right-click the selected rectangle, point to Order, and click Send to Back.

Finally, you'll center the title within the rectangle. Because ArcMap aligns elements in relation to the last selected element, you'll select the title first.

22 Click the title. Make sure the title, not the rectangle, is selected.

23 Hold down the Shift key and click the bottom of the rectangle, outside the title's selection box.

Both the grouped title and the rectangle should be selected. The rectangle's selection color is blue, indicating that it is the last selected element. The title's selection color changes to green.

24 Right-click either selected element, point to Align, and click Align Vertical Center.

25 Right-click either selected element, point to Align, and click Align Center.

The title is centered vertically and horizontally within the rectangle.

18

19

26 Right-click either selected element once more and click Group, then click an empty part of the virtual page to unselect the grouped element.

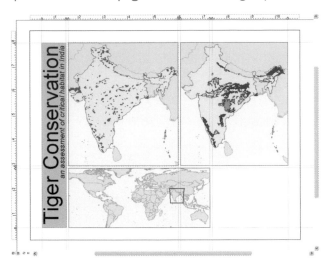

In the next exercise, you'll add a north arrow, scale bar, and legend to the Proposed reserves data frame.

27 If you want to save your work, save it as **my_ex19b.mxd** in the **\GTKArcGIS\Chapter19 \MyData** folder.

28 In the ArcMap application title bar, click the middle button in the upper right corner to restore the application window to its former size.

29 If you are continuing with the next exercise, leave ArcMap open. Otherwise, exit the application. Click No if prompted to save your changes.

Adding a north arrow, scale bar, and legend

North arrows, scale bars, and legends are associated with the data frame that is active when they are inserted. A scale bar changes when the display scale of its data frame changes. A legend is updated when layers in its data frame are deleted or resymbolized. Although these elements can be moved anywhere on a map layout, it is usually best to keep them within the data frame they belong to.

ArcMap has many different styles for north arrows, scale bars, and legends and allows you to customize them.

Exercise 19c

The people who review the proposal and your map will be familiar with the geography of India. Still, legends are necessary for interpreting any map, while scale bars and north arrows provide geographic orientation.

1   In ArcMap, open **ex19c.mxd** from the **C:\ESRIPress\GTKArcGIS\Chapter19** folder.

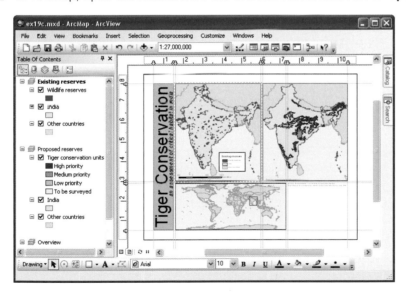

The map opens in layout view. It looks as it did at the end of the previous exercise, except that some additional elements (such as the legend and scale bar in the Existing reserves data frame) have been added for you. You will add more elements to the map, beginning with a north arrow in the Overview data frame.

2   In the ArcMap application title bar, click the middle button in the upper right corner to maximize the application window.

3   On the Layout toolbar, click the Zoom Whole Page button.

4   In the display area, click the Overview data frame to select it.

5   Click the Insert menu and click North Arrow. In the dialog box, click ESRI North 3, as shown in the following graphic, and click OK.

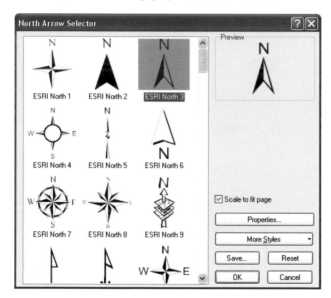

The north arrow is added to the map.

6   Drag the north arrow to the lower left corner of the Overview data frame. Make the north arrow smaller by dragging one of its selection handles, then adjust its position to leave a little space beneath it. (Use the following graphic as a guide.) When you're finished, click an empty part of the virtual page to unselect the north arrow.

Now you'll add a scale bar to the Proposed reserves data frame. Scale bars tell you how to convert measurements on the map into the real distances they represent on the ground.

7 Click the Proposed reserves data frame to select it.

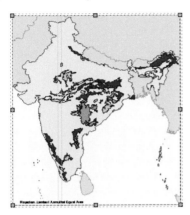

8 Click the Insert menu, and click Scale Bar.

9 In the Scale Bar Selector dialog box, click Alternating Scale Bar 1, as shown in the following graphic, and click OK.

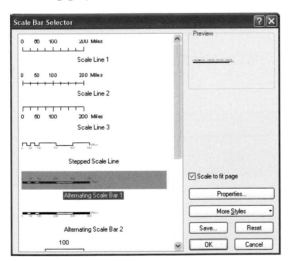

The scale bar is added to the layout. Before moving it into position, you'll change some of its properties to match the scale bar in the Existing reserves data frame.

10 Right-click the scale bar and click Properties. In the Alternating Scale Bar Properties dialog box, click the Scale and Units tab if necessary.

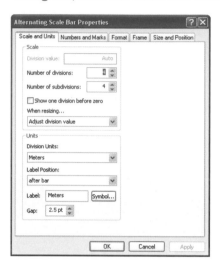

The scale bar currently has four divisions. The left-most division itself has four subdivisions. The scale bar units are meters.

11 In the middle of the dialog box, click the When resizing drop-down arrow and click Adjust width.

This will make the width of the scale bar change if the display scale of the data frame changes.

Adjusting scale bars

A scale bar changes as you zoom in or out on its data frame. If the scale bar is adjusted by division value, the distance represented by a scale bar division is variable. As you zoom in, it represents less distance; as you zoom out, it represents more. If the scale bar is adjusted by the number of divisions, the distance represented by a division stays the same, but there are fewer divisions as you zoom in and more as you zoom out. If the scale bar is adjusted by width, the distance represented by a division and the number of divisions stays the same, but the entire scale bar grows wider as you zoom in and narrower as you zoom out. (If a scale bar is adjusted by width, you cannot widen it by dragging; its width changes only with the display scale.)

12 In the Division value box at the top of the dialog box, replace the current value with **500**.

13 Click the Number of divisions down arrow and change the value to 3. Click the Number of subdivisions down arrow and change its value to 2.

14 Check Show one division before zero.

15 In the lower portion of the dialog box, click the Division Units drop-down arrow. Scroll up and click Kilometers. Make sure that your dialog box matches the following graphic, then click OK.

In the layout, the scale bar reflects the new properties you set.

16 Drag the scale bar to the lower left corner of the Proposed reserves data frame, just above the line of text.

You'll change its height to match the other scale bar. (You can't change the scale bar's width because of the resizing choice you made in step 11. But you don't need to change its width—as long as the two data frames have the same display scale, their scale bar widths will also be the same.)

18
19

17 With the scale bar selected, hold down the Shift key and click the scale bar in the Existing reserves data frame.

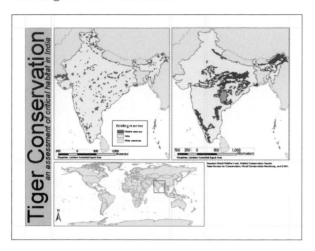

18 Both scale bars are selected. The Existing reserves scale bar, selected last, is outlined in blue.

19 Right-click either scale bar. On the context menu, point to Distribute and click Make Same Height. Right-click again on either scale bar. On the context menu, point to Align and click Align Bottom.

The two scale bars now look the same and have the same property settings. Next, you'll add a legend to the Proposed reserves data frame.

20 Click the Proposed reserves data frame to select it. Click the Insert menu and click Legend to open the Legend Wizard.

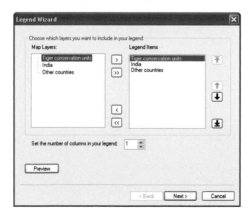

By default, the legend includes all layers from the map and the number of legend columns is set to one.

21 Click Next to accept the defaults.

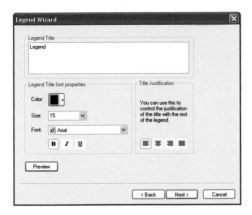

In the second panel you choose a title and style properties for the legend.

22 Replace the default title with **Proposed reserves**, as shown in the following graphic. Click Next.

In the third panel, you'll add a border to the legend.

23 Click the Border drop-down arrow and click 1.0 Point. Click the Background drop-down arrow, scroll to the bottom of the list, and click White.

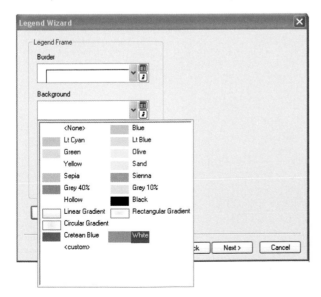

18

19

The next two panels set symbols and spacing for the legend, but you will not change these. Clicking Preview at any time displays the legend on the map in its current form and allows you to exit the wizard or return and make further changes.

24 On the wizard panel, click Preview. Move the wizard away from the map.

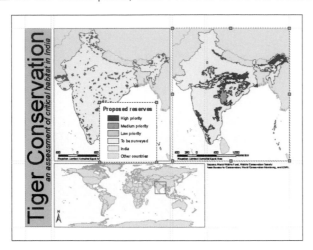

The legend is added to the map and the wizard remains open.

25 On the wizard panel, click Finish.

Finally, you'll resize and move the legend.

26 Drag the legend somewhere over the Proposed reserves data frame. Place the mouse pointer over any selection handle, drag the legend to about half its original size, and position it as shown in the following graphic. When you're finished, click an empty part of the virtual page to unselect the legend.

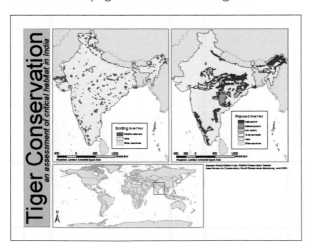

In the next exercise, you'll add a few more map elements and prepare to print your map.

27 If you want to save your work, save it as **my_ex19c.mxd** in the **\GTKArcGIS\Chapter19 \MyData** folder.

28 In the ArcMap application title bar, click the middle button in the upper right corner to restore the application window to its former size.

29 If you are continuing with the next exercise, leave ArcMap open. Otherwise, exit the application. Click No if prompted to save your changes.

18

19

Adding final touches and setting print options

Before sending a map to print, you should check your page setup options and preview the map in ArcMap.

Exercise 19d

The map is almost ready. You'll add a picture and a neatline, preview the map, and send it to a printer.

1   In ArcMap, open **ex19d.mxd** from the **C:\ESRIPress\GTKArcGIS\Chapter19** folder.

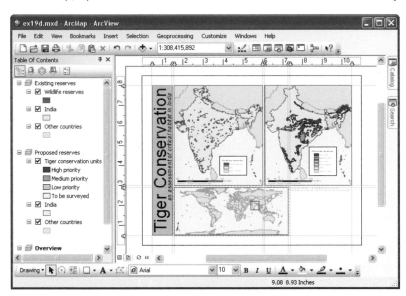

The map opens in layout view. It looks as it did at the end of the previous exercise.

2   In the ArcMap application title bar, click the middle button in the upper right corner to maximize the application.

3   On the Layout toolbar, click the Zoom Whole Page button.

4   Click the Insert menu and click Text.

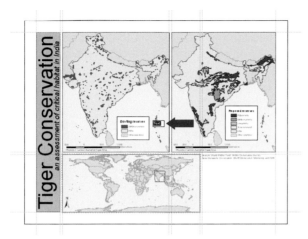

The text box appears in the middle of the layout with the word "Text" highlighted.

5   In the text box, type **Projection: Geographic** and press Enter. The text is selected.

6   On the Draw toolbar, highlight the value in the Font Size drop-down list and replace it with **6.5**. Press Enter.

7   Drag the text to the lower left corner of the Overview data frame and place it beneath the north arrow. Unselect the text.

You'll add a photograph of a tiger. Who could resist a picture of such a magnificent animal?

8   Click the Insert menu and click Picture.

9   In the Open dialog box, navigate to **\GTKArcGIS\Chapter19\Data**. Click **tiger.jpg** and click Open.

The tiger image is added to the map.

18

19

10 Right-click an empty part of the virtual page. On the context menu, point to Guides and click Snap to Guides.

11 Place the mouse pointer over the image and drag it so that its lower right corner snaps to the guides at 0.5 inches (vertical ruler) and 10.5 inches (horizontal ruler).

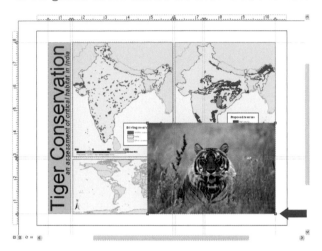

12 Place the mouse pointer over the selection handle in the upper left corner of the image. Drag to resize the image so that its left edge snaps to the guide at 7.4 inches (horizontal ruler). Make sure the tiger image does not hide the text that is positioned above it. Unselect the image.

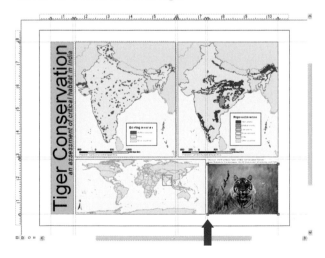

The map is almost finished. The last element you add will be a neatline, a bounding line that frames the other elements.

13 Click the Insert menu and click Neatline.

14 In the Neatline dialog box, choose the option to place the neatline around all elements. Click the Border drop-down arrow and click 2.5 Point.

15 Click the Background drop-down arrow and click None. Make sure that your dialog box matches the following graphic, then click OK.

16 Unselect the neatline.

Before you print a map, you should make sure that your map size and printer setup specifications match. Otherwise, your map may be cut off or misaligned.

17 Click the File menu and click Page and Print Setup.

In exercise 19a, you changed the map page orientation from Portrait to Landscape. Now you'll change your paper orientation, too.

18 In the Paper frame of the Page and Print Setup dialog box, click the Landscape option. If necessary, click the Name drop-down arrow and click the printer you want to use. Accept the default paper size (Letter). Make sure that your dialog box matches the following graphic (except for the printer name). Click OK.

18
19

You are ready to preview your map. The preview shows you how the map will look on the printed page, so you can correct any mistakes in advance.

19 Click the File menu and click Print Preview.

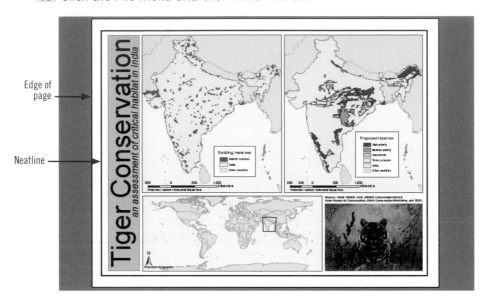

At this point, you should be mainly concerned with the alignment of elements on the page. Your alignment will depend on the type of printer you are connected to and its configuration. If the map elements overlap the page edge, you'll correct this problem in a moment.

Don't worry that the quality of the tiger image is poor in the preview—it will be fine when it prints.

20 Click Print to open the Print dialog box.

21 If the map looked good in the print preview, click OK to print it.

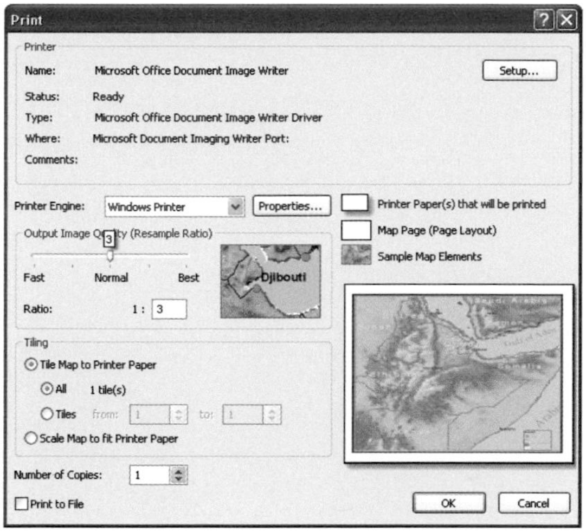

22 If the map went over the edge of the page, click the Scale Map to fit Printer Paper option in the lower left corner of the dialog box. Click OK.

You have finished the map.

23 If you want to save your work, save it as **my_ex19d.mxd** in the **\GTKArcGIS\Chapter19 \MyData** folder.

24 In the ArcMap application title bar, click the middle button in the upper right corner to restore the application window to its former size.

In the next chapter, you'll launch ArcMap from ArcCatalog. So even if you are continuing, you should exit ArcMap.

25 Close ArcMap. Click No if prompted to save your changes.

18

19

Chapter 20

Creating models

Starting a model
Building a model
Enhancing a model

GIS is often defined as a computerized system for the creation, management, query, analysis, and display of spatial data. The organization of this book reflects that task-oriented definition. But GIS is also defined as a decision support system, in which computer software is used to model spatial processes and to solve problems analytically. This process-oriented definition applies to the book's final chapter.

In chapters 11 and 12, during the tree-harvesting analysis, you learned how to use some of the geoprocessing tools in ArcToolbox, such as Dissolve, Clip, Buffer, and Union. Individually, these tools are useful for completing particular tasks, but their real power, as you saw, lies in their being used together to solve a complex problem.

The tools in a real toolbox (hammers, saws, screwdrivers) have their own special uses, too. But again, if you want to accomplish something significant—like building a house—you have to use these tools together in a carefully planned sequence of steps.

Carpenters build a house from a blueprint or scale model. In ArcGIS, you can use Model-Builder to make a blueprint of your analysis. Like a good carpenter, ArcGIS will then carry out the individual procedures in order according to your plan.

ModelBuilder is a graphical interface for diagramming solutions to spatial analysis problems. The diagrams you make are called models. Below, a very simple model shows a Buffer operation being applied to an input dataset to create a new output dataset.

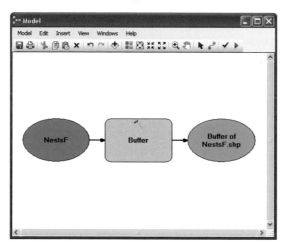

A complete model is an interconnected set of such processes, each consisting of input data, a tool, and the resulting output data. The next model consists of three processes: two buffers and a union. Notice that the outputs of the buffer processes become the inputs to the union process.

The elements of a process are distinguished by their symbology. By default, input data is represented by a blue oval. Tools are represented by yellow rectangles. Output data is represented by a green oval. The representative shapes can be switched if you wish.

Once you have diagrammed a model, you can run it with a single click. ModelBuilder runs each of the processes in turn, passing the output of one process on to the next process until a final dataset is produced.

Suppose you were asked to re-create the analysis you did in chapters 11 and 12, but with a slight variation, such as a different goshawk nest buffer? Or suppose you were asked to do exactly the same analysis, but for a different lease area?

You could do it, of course, by reworking the chapter exercises. You would open the tools in ArcToolbox, set their parameters, and run them—and it would take you just as long (or almost as long) as it did the first time. In fact, whenever you wanted to calculate harvest stand values, assuming this was a regular part of your job, you would have to go through the same lengthy process.

Now imagine that you have already modeled the analysis in ModelBuilder. When a new piece of information comes in—the protected zone around goshawk nests has been changed from 800 to 1,000 meters—you don't have to reconstruct your entire workflow. Instead, you just open the model, change a parameter in the nest buffer process, and rerun the model. That's all it takes. When the time comes to bid on a different lease, you can copy the model and change the data sources used as the initial inputs.

ModelBuilder can save you huge amounts of time if you perform repetitive spatial analyses. It also lets you evaluate the changing conditions and what-if scenarios that arise in decision-making situations.

Furthermore, models are a good way to evaluate the soundness of your methodology, to document this methodology for others, and to develop and refine it over time. Sharing models with colleagues helps to standardize decision-making processes. Of course, your timber company might not want to share its successful model with competitors, but if you worked for a government agency, you could use models to exchange knowledge and to establish uniform practices.

For more information about spatial modeling and ModelBuilder, click the Contents tab in ArcGIS Desktop Help and navigate to *Professional Library > Geoprocessing > Geoprocessing with ModelBuilder > What is ModelBuilder?*

Starting a model

ModelBuilder is not a free-standing software application like ArcMap or ArcCatalog, but it's more than just a geoprocessing tool: it's a design environment for creating workflow diagrams called models. Like tools, however, models are stored in a toolbox. In earlier chapters, you used some of the tools and toolboxes that come with ArcMap. In this exercise, you will learn how to create a new toolbox and put a model inside it.

A new toolbox can be created in ArcToolbox, or in an ordinary file folder, or in a geodatabase. Since models are stored in toolboxes, these are also the three locations in which models can be created.

The ModelBuilder window has its own menu and toolbar. Many of the controls have the same icons and functions as ArcMap controls.

Exercise 20a

In the next three exercises, you will create a model of the harvest profitability analysis from chapter 12. (The data preparation you did in chapter 11 is assumed to be complete, although you could include those operations in your model as well.)

The logic of the analysis is not repeated in detail here. If you feel unsure about the order of the steps or the reasons behind them, you can refer to chapter 12. The focus of this chapter is mainly on how to diagram the analysis in ModelBuilder.

In this exercise, you'll make a new toolbox to store your model. Then you'll create a model and add one process to it.

1   Start ArcCatalog. In the catalog tree, navigate to **C:\ESRIPress\GTKArcGIS\Chapter20 \MyToolboxes**. Right-click **MyToolboxes**, point to New, and click Toolbox.

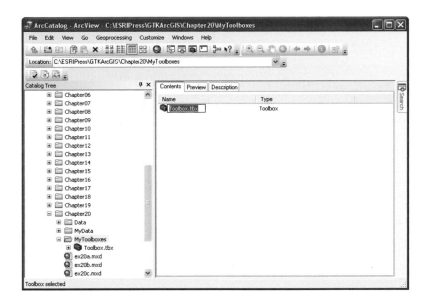

A new toolbox is added with its default name highlighted.

2 In the catalog display, change the new toolbox name to **Harvest Models** and press Enter. In the catalog tree, expand the MyToolboxes folder if necessary.

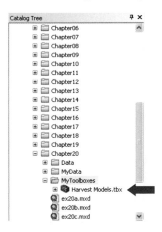

You have created a toolbox in an ordinary file folder. If you look at the toolbox with a file browser, such as Windows Explorer, it appears with a .tbx extension. (Depending on your setup, you may have file extensions hidden. To change how file extensions are viewed in ArcCatalog, click Customize, click ArcCatalog Options, and check or uncheck the Hide file extensions check box.)

Now you will create a model within the toolbox.

20

3   In the catalog tree, right-click the Harvest Models toolbox, point to New, and click Model.

The model window opens. You'll change the name in the title bar to something more meaningful.

4   In ModelBuilder, click the Model menu and click Model Properties. In the Model Properties dialog box, the General tab is selected. In the Name box, replace Model with **HarvestProfitability** (the model name cannot contain any spaces). In the Label box, replace Model with **Harvest Profitability** (the model label may contain spaces). Make sure the dialog box matches the following graphic, then click OK.

The title bar is updated with the new name. Now you will save and close the model so you can work with it in ArcMap.

5   On the ModelBuilder toolbar, click the Save button.

6 Click the Model menu and click Close.

7 In the catalog tree, click the plus sign next to the Harvest Models toolbox.

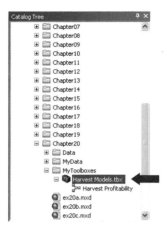

The new toolbox is not automatically added to new ArcMap documents. To work with the Harvest Profitability model, you must first add the Harvest Models toolbox to ArcToolbox from within ArcMap.

8 In the catalog tree, double-click **ex20a.mxd** to launch ArcMap and open the map document. (If ArcMap is already running, click the File menu and click Open. Navigate to **\GTKArcGIS\Chapter20**. Click **ex20a.mxd** and click Open.)

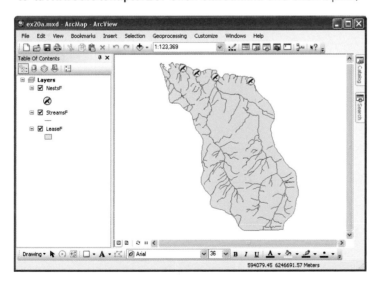

The map contains layers for lease F, goshawk nests, and streams. It should look familiar as it has the same data and symbology as ex12a.mxd.

Now you will open the ArcToolbox window and add the Harvest Models toolbox to it.

9 On the Standard toolbar, click the ArcToolbox window button to open ArcToolbox. (Depending on how you previously setup your ArcToolbox window, it may be floating, docked or tabbed along the side of the main window.)

10 Position ArcMap and ArcCatalog so that the ArcToolbox window and the catalog tree don't overlap. Make ArcCatalog the active application. In the catalog tree, click the Harvest Models toolbox and drag it to the bottom of the ArcToolbox window.

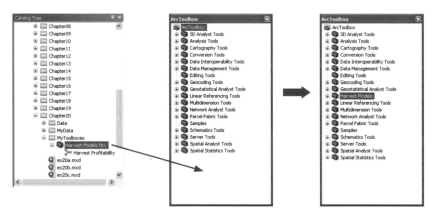

The Harvest Models toolbox is automatically put in alphabetical order with the other toolboxes.

11 Make ArcCatalog the active application. Click the File menu and click Exit.

12 In ArcMap, in the ArcToolbox window, click the plus sign next to Harvest Models to expand it. Right-click the Harvest Profitability model and click Edit. When the ModelBuilder window opens, move it so that it doesn't overlap either the ArcToolbox window or the ArcMap table of contents.

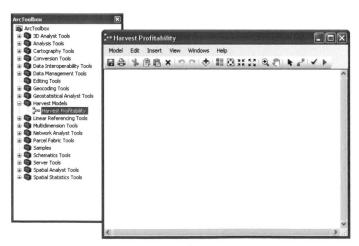

Now you will create a process to buffer the goshawk nests. You will drag and drop the Buffer tool from ArcToolbox and the NestsF layer from the ArcMap table of contents.

13 In the ArcToolbox window, click the plus sign next to Analysis Tools, then click the plus sign next to Proximity.

14 Click the Buffer tool and drag it to ModelBuilder. Drop it in the center of the ModelBuilder window.

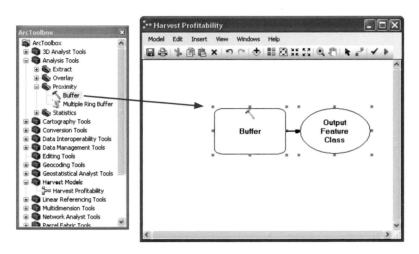

The Buffer tool, represented by a rectangle, is added to the model. An arrow connects the tool to its output data, represented by an oval. Blue selection handles indicate that the elements are currently selected. The fact that they are not colored in means that they are incomplete—they are waiting for input data to be attached.

15 In the ArcMap table of contents, click the NestsF layer. Drag it to the ModelBuilder window and drop it directly on top of the Buffer tool rectangle. A context menu appears, click Input Features.

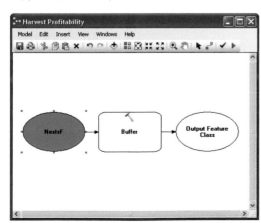

The NestsF layer, represented by an oval, is automatically connected to the tool it was dropped onto.

Now you will change the tool and output data settings.

16 In the model, double-click the Buffer tool rectangle.

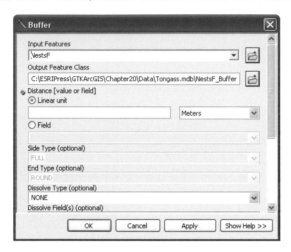

The Buffer dialog box opens—exactly as if you had double-clicked the Buffer tool in ArcToolbox. You will fill in the tool's parameters just as you did in chapter 12.

By default, the output feature class is set to the Tongass.mdb file in the chapter 20 Data folder. To keep the model data separate from the book's exercise data, you will change the output location to the MyData folder.

17 Click the Browse button next to the Output Feature Class box. In the Output Feature Class dialog box, navigate to **\GTKArcGIS\Chapter20\MyData** and double-click **MyTongass.mdb**.

18 In the Name box, type **NestBuf**, then click Save.

The output feature class information is updated in the Buffer dialog box.

19 For the buffer distance, make sure the Linear unit option is selected. In the Linear unit box, type **800**.

The distance units are correctly set to meters. Trees cannot be harvested within 800 meters of a goshawk nest.

20 Click the Dissolve Type drop-down arrow and click All.

Overlapping buffer polygons will be dissolved to make a single feature.

21 Make sure the dialog box matches the following graphic, then click OK to close the tool.

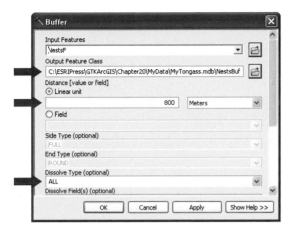

Clicking OK here does not run the tool, it just closes the dialog box. In ModelBuilder, you can double-click a tool whenever you like to view or change its parameters.

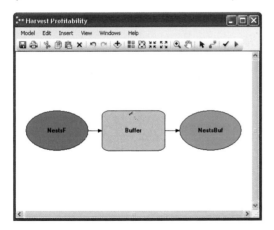

The elements are colored in now, which means the process is ready to run.

When you run the process, a new feature class will be created and saved to disk. The output data is not automatically added as a layer to ArcMap, however. If you want the data added to the map, you have to set a property on the output data element.

22 In the model, right-click the NestBuf oval and click Add To Display to check it on.

23 On the ModelBuilder toolbar, click the Run button to run the model.

20

As a model runs, each of its tools turns red. Your model has only one tool, so if you weren't watching closely (or if the progress report covered the model), you may not have seen it. When the tool is finished, it turns yellow again.

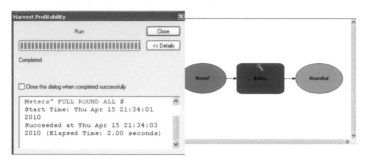

24 Click Close to close the progress report. Right click NestBuf and select Add to display.

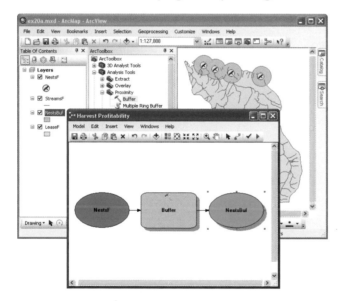

The NestBuf layer is added to ArcMap. (You may need to move the ModelBuilder window to see it.)

In ModelBuilder, drop shadows behind the tool and output data elements indicate that the process has been run. Once a process has been run, it does not run again unless you change a parameter, or choose Run Entire Model or Validate Entire Model from the Model menu.

25 On the ModelBuilder toolbar, click the Save button to save the model.

26 Click the Model menu and click Close to close the model. In ArcMap, close the ArcToolbox window.

The model and all its settings are stored in the Harvest Models.tbx file you created at the beginning of this exercise. There is no need to save changes to the map document.

27 If you are continuing with the next exercise, leave ArcMap open. Otherwise, exit the application. Click No if prompted to save your changes.

20

Building a model

Most models consist of several connected processes. To connect processes—that is, to define the output of one process as the input to another—you draw connector lines between them with the Add Connection tool.

Exercise 20b

In this exercise, you will add a second buffer process for streams. Then you will add a Union process that overlays the stream and nest buffers to define the area where trees cannot be harvested (the "no-cut" zone). Finally, you will union the no-cut zone with the original layer of forest stands. The output of this union will be a dataset in which forest stands are split wherever they cross the no-cut zone—in other words, harvestable land will be geometrically separate from unharvestable land.

1   In ArcMap, open **ex20b.mxd** from the **C:\ESRIPress\GTKArcGIS\Chapter20** folder.

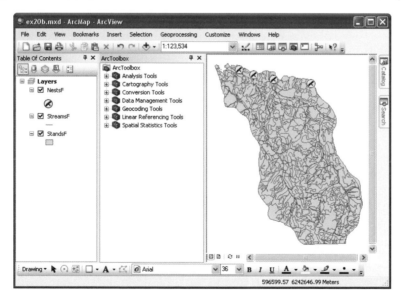

In the map you see the nests, streams, and stands of lease F.

2   If needed, click the ArcToolbox window button to open ArcToolbox.

In the last exercise, you dragged the Harvest Models toolbox from ArcCatalog to ArcToolbox. You can also add toolboxes from within ArcMap.

3   In the ArcToolbox window, right-click a blank area and click Add Toolbox. In the Add Toolbox dialog box, navigate to **\GTKArcGIS\Chapter20\MyToolboxes**. Click **Harvest Models** and click Open.

4   In ArcToolbox, click the plus sign next to Harvest Models to expand it. Right-click Harvest Profitability and click Edit to open the model. Move the ModelBuilder window so that it doesn't overlap the ArcMap table of contents or the ArcToolbox window.

5   In ArcToolbox, click the plus sign next to Analysis Tools, then click the plus sign next to Proximity. Click the Buffer tool, then drag and drop it at the top center of the ModelBuilder window, as shown in the following graphic.

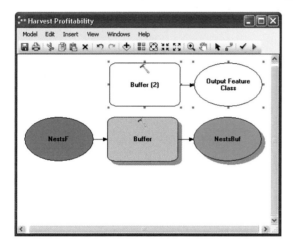

Don't worry about its exact position. In a few steps, you will have ModelBuilder tidy up.

6   In the ArcMap table of contents, click the StreamsF layer. Drag and drop it directly on top of the Buffer (2) tool. In the context menu that appears, click Input Features.

When you drop the layer, the input data is automatically connected to the tool as an input feature. Before the model is ready to run, however, you need to change the tool's settings.

7   Double-click the Buffer (2) tool.

8   Click the Browse button next to the Output Feature Class box. In the Output Feature Class dialog box, navigate to **\GTKArcGIS\Chapter20\MyData** and double-click **MyTongass.mdb**. In the Name box, type **StreamBuf**, then click Save.

You may recall from chapter 12 that logging is prohibited within 50 meters of streams, and within 100 meters of streams where salmon spawn. The StreamsF layer attribute table has a field called Distance, which contains values of either 50 or 100 for each stream feature. You want the buffer size for each stream feature to be determined by the values in this field.

20

9 For the buffer distance, click the Field option. Click the Field drop-down arrow and click Distance. Make sure that your dialog box matches the following graphic, then click OK.

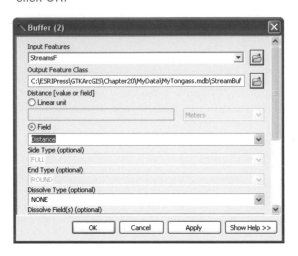

Your model now has two processes and will soon have more. Aligning elements to keep the model looking neat could be a very time-consuming business. Fortunately, ModelBuilder has an Auto Layout button that takes care of this for you.

10 In ModelBuilder, click the Auto Layout button, then click the Full Extent button.

The processes are moved and aligned.

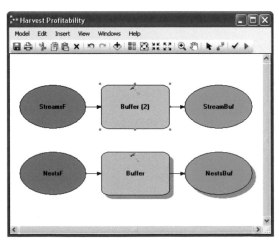

Before adding a third process, you will zoom out and pan the model.

11 Click the Fixed Zoom Out button once, then click the Pan tool.

12 Move the two processes to the left side of the ModelBuilder window, as shown in the following graphic.

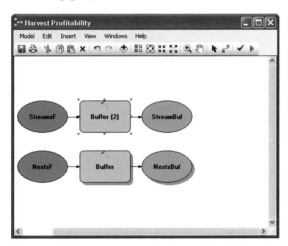

NestBuf and StreamBuf together define the area in which trees cannot be harvested. To make a single no-cut zone, you will add a Union process to overlay these two buffer layers.

13 In the ArcToolbox window, click the plus sign next to Overlay (in the Analysis Tools toolbox). Click the Union tool. Drag and drop it on the right side of the model, between NestBuf and StreamBuf, as shown in the following graphic.

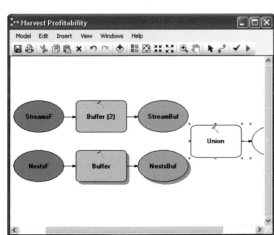

20

Don't worry that the output dataset is only partly visible. You'll fix this later with the Auto Layout tool.

The Union tool is not yet colored in because it has no input data. To define the two buffer datasets as the inputs, you will add connector lines. (Previously, you dropped input data directly on the buffer tools and ModelBuilder made the connections for you.)

14 Click the Add Connection tool.

15 In the ModelBuilder window, notice that the cursor changes (it looks like a magic wand.) Click the green NestBuf oval and then click the Union tool. In the context menu that appears, click Input Features.

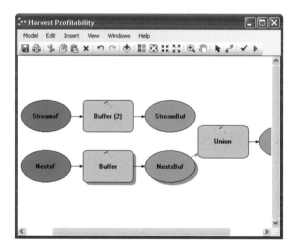

A cyan connector line points from NestBuf to Union, and the process is technically ready to run. A Union overlay on a single dataset would be pointless, however, since the output would be identical to the input.

16 Click the green StreamBuf oval, then click the Union tool to add a second connector line. Click Input features.

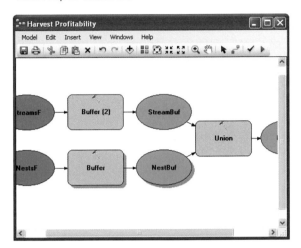

The new connector line is cyan, and the one you added before turns black. Now you will arrange the model and zoom out to see all its elements.

17 Click the Auto Layout button, then click the Full Extent button.

You see the entire model, although you may not be able to read the labels. Enlarge the ModelBuilder window if you like, and again click the Auto Layout and Full Extent buttons.

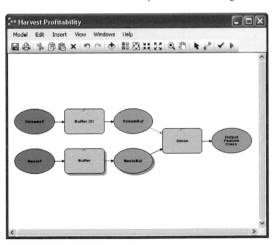

20

18 Double-click the Union tool.

In the Union dialog box, NestBuf and StreamBuf are the input features (as you specified). You will change the name of the output feature class.

19 Click the Browse button next to the Output Feature Class box. In the Output Feature Class dialog box, navigate, if necessary, to **\GTKArcGIS\Chapter20\MyData** and double-click **MyTongass.mdb**. In the Name box, type **NoCutArea**. Click Save.

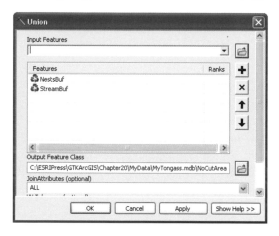

20 Click OK to close the tool.

The NoCutArea layer marks the territory that is off-limits to tree harvesting. But the NoCutArea polygons are simply zones drawn around nest and stream features—they don't tell you which forest stands can or cannot be harvested. To find out, you must overlay the no-cut area with a layer of forest stands for Lease F.

This overlay will split forest stand polygons wherever they cross NoCutArea polygons. In the output dataset, every polygon will lie either completely inside the no-cut area or completely outside it. (Stands that do not cross the no-cut area don't have to be split.)

After the overlay, you will have a layer of forest stands in which every polygon has an attribute that identifies it as harvestable or not. You will then go on to recalculate the values of the harvestable stands in the next exercise.

21 In the ArcToolbox window, click the Union tool. Drag and drop it underneath the existing Union tool, as shown in the following graphic.

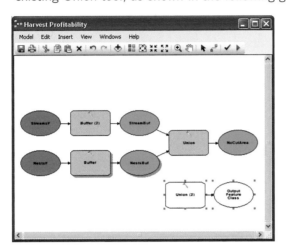

22 In the ArcMap table of contents, click the StandsF layer. Drag and drop it on top of the Union (2) tool. In the context menu that appears, click Input Features.

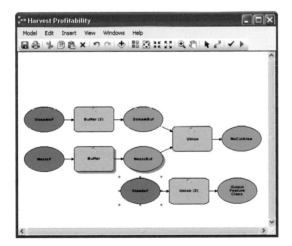

ModelBuilder connects the process automatically. Like the previous Union process, this one is now ready to run even though it has only a single input. To complete the overlay, you will connect the NoCutArea oval to the Union (2) tool.

20

23 Make sure the Add Connection tool is still active. Click the green NoCutArea oval, then click the Union (2) tool to add a connector line. In the context menu that appears, click Input Features.

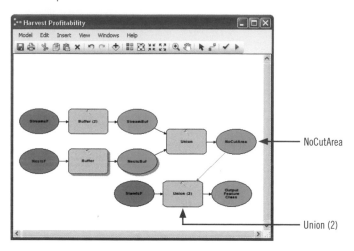

24 On the ModelBuilder toolbar, click Auto Layout, then click Full Extent.

The model is rearranged.

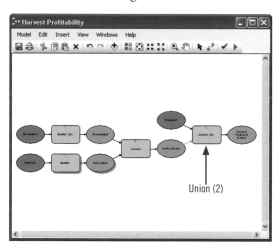

25 Double-click the Union (2) tool.

26 In the Union (2) dialog box, click the Browse button next to the Output Feature Class box. In the Output Feature Class dialog box, navigate, if necessary, to **\GTKArcGIS \Chapter20\MyData** and double-click **MyTongass.mdb**. In the Name box, type **NoCutAndStands**, then click Save.

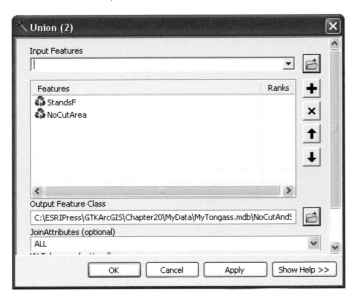

27 Click OK to close the tool.

28 In the model, right-click the green NoCutAndStands oval (the element farthest to the right in the model) and click Add To Display.

The model now has four processes—two buffers and two overlays—and is ready to run. The nest buffer process, which ran in the previous exercise, will not run again. (It still has drop shadows under the tool and output data elements.)

29 Click the Run button to run the model.

In the processes that have not been run, the tools turn red as they execute.

20

30 When the model is done, click Close on the progress report.

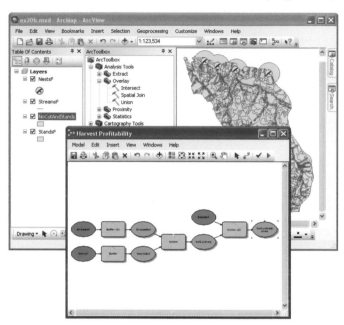

In the ModelBuilder window, all processes are drop-shadowed. In ArcMap, the NoCutAndStands layer is added to the map.

31 On the ModelBuilder toolbar, click the Save button.

32 Close the model.

33 Close the ArcToolbox window.

Your edits to the model are saved to the model's toolbox file when you click the Save button in ModelBuilder. You don't need to save changes to the map document.

34 If you are continuing with the next exercise, leave ArcMap open. Otherwise, exit the application. Click No if prompted to save your changes.

Enhancing a model

So far, your model has made use of two analysis tools, Buffer and Union. ModelBuilder lets you use many other kinds of tools as well. The Data Management toolbox contains the Make Feature Layer tool, which creates an ArcMap selection layer from a query, and the Calculate Field tool, which uses a logical or mathematical expression to change record values for a field in a table.

Exercise 20c

In this exercise, you will complete the Harvest Profitability model. First, you will use the Make New Layer tool to query for harvestable stands and to create a selection layer from them.

You may recall from chapter 12 that when two layers are overlaid, the output layer has the attributes of both input layers, including their FID fields. If an output feature spatially corresponds to an input feature from Layer X, it gets that feature's identifier in its FID_Layer X field. Otherwise, it gets the value -1 in this field instead.

The previous output from your model was the NoCutAndStands layer. It has FID attributes from the two layers it unioned: NoCutArea and StandsF. The important attribute for your analysis is FID_NoCutArea. If a NoCutAndStands polygon has the value -1 in this field, it means that the polygon does not spatially correspond to a feature from NoCutArea; in other words, the polygon is harvestable. This is the vital piece of information you need to construct your query.

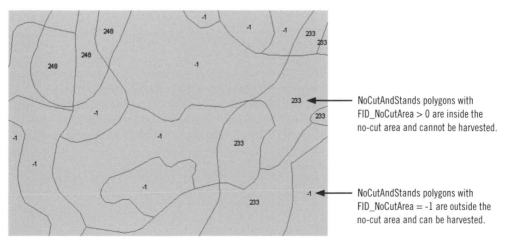

NoCutAndStands polygons with FID_NoCutArea > 0 are inside the no-cut area and cannot be harvested.

NoCutAndStands polygons with FID_NoCutArea = -1 are outside the no-cut area and can be harvested.

20

After selecting harvestable stands, you will use the Calculate Field tool to recalculate their values. Many of these stands are new polygons that were split during the union of the StandsF and NoCutArea layers in the last exercise. These new polygons have correct area values, because ArcMap updated their Shape_Area attribute automatically. They have incorrect StandValue attributes, however. StandValue is calculated by multiplying Shape_Area and ValuePerMeter, but ArcMap does not update calculated fields on its own.

1   In ArcMap, open **ex20c.mxd** from the **C:\ESRIPress\GTKArcGIS\Chapter20** folder.

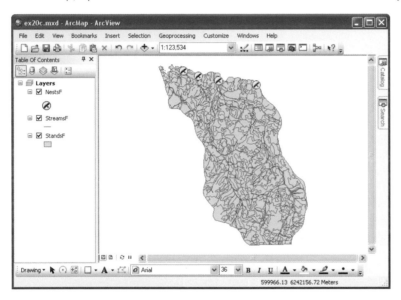

In the map you see the nests, streams, and stands of lease F.

Once again, you will add the Harvest Models toolbox to ArcToolbox.

2   Click the ArcToolbox window button. In ArcToolbox, right-click a blank area and click Add Toolbox. In the Add Toolbox dialog box, navigate to **\GTKArcGIS\Chapter20 \MyToolboxes**. Click **Harvest Models** and click Open.

3   In ArcToolbox, click the plus sign next to Harvest Models to expand it. Right-click the Harvest Profitability model and click Edit.

The model looks as it did at the end of the previous exercise.

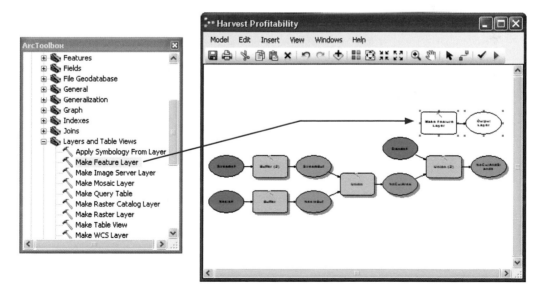

4 Position the model so that it doesn't overlap ArcToolbox or the ArcMap table of contents.

You'll start by adding the Make Feature Layer tool.

5 In ArcToolbox, click the plus sign next to the Data Management Tools toolbox. Click the plus sign next to the Layers and Table Views tools. Click the Make Feature Layer tool, then drag and drop it to the upper right corner of the model, as shown in the following graphic.

20

6   On the ModelBuilder toolbar, click the Add Connection tool.

7   Click the NoCutAndStands green oval, then click the Make Feature Layer tool to connect them. In the context menu that appears, click Input Features.

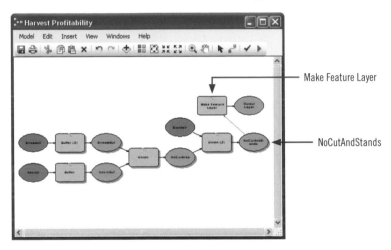

8   Click the Auto Layout button, then click the Full Extent button.

The model is rearranged.

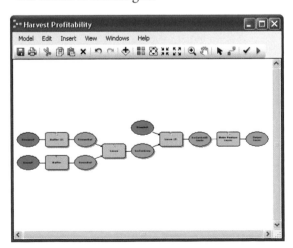

9 Double-click the Make Feature Layer tool. (It is now the yellow rectangle farthest to the right.)

You will write an SQL query to select harvestable stands and make a selection layer from them. This operation does not create a new dataset. The selection layer is based on the NoCutAndStands dataset that already exists. (NoCutAndStands is the last actual dataset that the model will produce.)

10 In the Output Layer box, replace the default name with **Harvest Stands**.

11 Next to the Expression box, click the SQL button to open the Query Builder dialog box.

To select stands outside the no-cut area, you will create the expression [FID_NoCutArea] = -1.

12 In the Fields scrolling box, double-click [FID_NoCutArea] to add it to the expression box. Click the equals (=) button. Press the space bar, and type -1.

13 Make sure that your expression matches the following graphic, then click OK.

14 Click OK on the Make Feature Layer dialog box.

Next, you will recalculate the stand value of the harvestable stands. Some of these stands (the ones that weren't split during the last overlay) still have correct stand values, but many do not.

15 In the ArcToolbox window, underneath Data Management Tools, click the plus sign next to Fields. Click the Calculate Field tool. Drag and drop the tool in the upper right corner of the model, as shown in the following graphic.

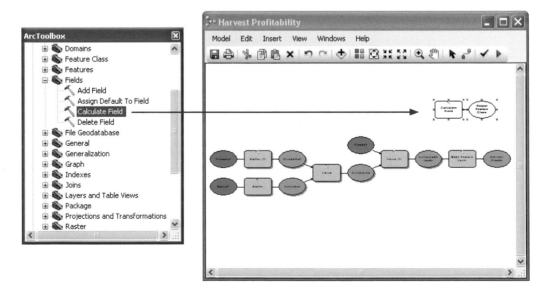

16 Make sure the Add Connection tool is still active. Click the green Harvest Stands oval, then click the Calculate Field tool to add a connector line between them. In the context menu that appears, click Input Table.

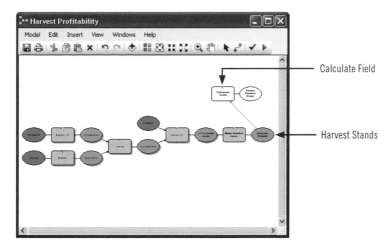

17 Click the Auto Layout button, then click the Full Extent button.

The Calculate Field tool is not ready to run because the tool does not yet have an expression. You will add that next.

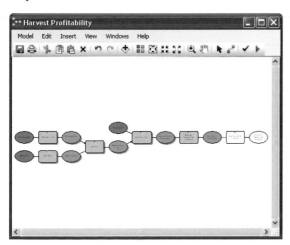

20

18 Double-click the Calculate Field tool (the white rectangle) to open the Calculate Field dialog box.

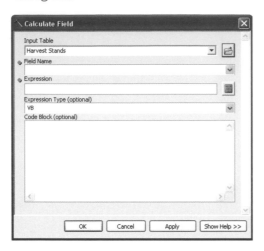

The input table is set to Harvest Stands. (This is the connection you just drew yourself.)

19 Click the Field Name drop-down arrow and click StandValue.

20 Next to the Expression box, click the Calculator button to open the Field Calculator.

21 In the Fields scrolling box, double-click Shape_Area to add it to the expression box. Click the multiplication (*) button. In the Fields scrolling box, double-click ValuePerMeter.

This expression will give you the updated stand values in dollars. In the table, however, the stand values are expressed in millions of dollars.

22 Click at the beginning of the expression and type an open parenthesis "**(**" followed by a space. Click at the end of the expression and type a space followed by a close parenthesis "**)**". Click the division (/) button, then type a space and type **1000000**. Make sure your expression matches the one in the following graphic, then click OK on the Field Calculator.

The Calculate Field dialog box is updated with the expression.

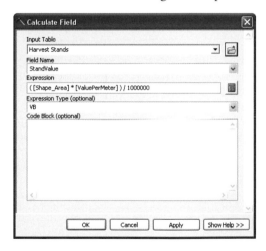

23 Click OK to close the tool.

Now that the expression has been entered, the process is colored in. You are ready to run the model. (Again, only processes that haven't run before will run.)

24 Click the Run button to run the model.

20

25 When the model is finished running, close the progress report.

You will add the Harvest Stands (2) layer to ArcMap and check the results of your calculation.

26 In the model, right-click the green Harvest Stands (2) oval and click Add To Display.

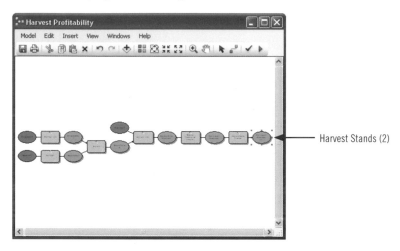

Harvest Stands (2)

Although Harvest Stands and Harvest Stands (2) are separate elements in the model, they represent the same selection layer. This layer is added to ArcMap.

27 In the ArcMap table of contents, right-click the Harvest Stands layer and click Open Attribute Table. Right-click the StandValue field name and click Statistics.

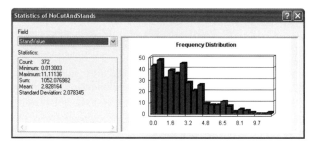

The sum of the recalculated values is 1,052.076982. Since the values are in millions, this is slightly over a billion dollars.

Note: You may have noticed ModelBuilder has given a slightly different sum than when you performed each step separately in chapter 12. This is due to rounding techniques, and the effect is negligible.

28 Close the statistics window. Close the Attributes of Harvest Stands table.

29 On the ModelBuilder toolbar, click the Save button to save the model.

30 Click the Model menu and click Close. In ArcMap, close the ArcToolbox window.

31 Click the File menu and click Exit. Click No when prompted to save your changes.

The model that you have created in this chapter is really just a part of a complete Harvest Profitability model. The model could be expanded to include the data preparation operations—like Dissolve and Clip—that you used in chapter 11. Another piece could analyze the costs of harvesting: crews have to be brought in to cut down the trees, load them on trucks, and remove them from the forest. Some of the analysis might be geospatial and some might be purely computational—you have seen that ModelBuilder can do both.

At the end of this exercise, you got the total harvestable stand value the old-fashioned way: by opening the table, right-clicking the field, and opening the Statistics window. With a little programming knowledge, you could write a script to do this using the Python programming language that comes with ArcGIS. (Scripts can be added to a model as tools.) When the model ran, the script would access the table, sum the stand values, and report the total in a pop-up window. Your model would then be fully automated.

As mentioned in the chapter introduction, ModelBuilder makes it easy to repeat an analysis on different project data. In this chapter, you calculated the harvest value for lease F, but you might be asked to do the same for another lease. The Data folder for chapter 20 contains a geodatabase called LeaseB.mdb with feature classes for the stands, nests, and streams of lease B. On your own, you might want to try adapting the Harvest Profitability model to work with this data.

If you decide to do that, you have a little more modeling ahead of you. If not, your work is done. You have come to the end of *Getting to Know ArcGIS Desktop*.

20

Data license agreement

Important: Read carefully before opening the sealed media package

ENVIRONMENTAL SYSTEMS RESEARCH INSTITUTE INC. (ESRI) IS WILLING TO LICENSE THE ENCLOSED DATA AND RELATED MATERIALS TO YOU ONLY UPON THE CONDITION THAT YOU ACCEPT ALL OF THE TERMS AND CONDITIONS CONTAINED IN THIS LICENSE AGREEMENT. PLEASE READ THE TERMS AND CONDITIONS CAREFULLY BEFORE OPENING THE SEALED MEDIA PACKAGE. BY OPENING THE SEALED MEDIA PACKAGE, YOU ARE INDICATING YOUR ACCEPTANCE OF THE ESRI LICENSE AGREEMENT. IF YOU DO NOT AGREE TO THE TERMS AND CONDITIONS AS STATED, THEN ESRI IS UNWILLING TO LICENSE THE DATA AND RELATED MATERIALS TO YOU. IN SUCH EVENT, YOU SHOULD RETURN THE MEDIA PACKAGE WITH THE SEAL UNBROKEN AND ALL OTHER COMPONENTS TO ESRI.

ESRI license agreement

This is a license agreement, and not an agreement for sale, between you (Licensee) and Environmental Systems Research Institute Inc. (ESRI). This ESRI License Agreement (Agreement) gives Licensee certain limited rights to use the data and related materials (Data and Related Materials). All rights not specifically granted in this Agreement are reserved to ESRI and its Licensors.

Reservation of Ownership and Grant of License: ESRI and its Licensors retain exclusive rights, title, and ownership to the copy of the Data and Related Materials licensed under this Agreement and, hereby, grant to Licensee a personal, nonexclusive, nontransferable, royalty-free, worldwide license to use the Data and Related Materials based on the terms and conditions of this Agreement. Licensee agrees to use reasonable effort to protect the Data and Related Materials from unauthorized use, reproduction, distribution, or publication.

Proprietary Rights and Copyright: Licensee acknowledges that the Data and Related Materials are proprietary and confidential property of ESRI and its Licensors and are protected by United States copyright laws and applicable international copyright treaties and/or conventions.

Permitted Uses: Licensee may install the Data and Related Materials onto permanent storage device(s) for Licensee's own internal use.

Licensee may make only one (1) copy of the original Data and Related Materials for archival purposes during the term of this Agreement unless the right to make additional copies is granted to Licensee in writing by ESRI.

Licensee may internally use the Data and Related Materials provided by ESRI for the stated purpose of GIS training and education.

Uses Not Permitted: Licensee shall not sell, rent, lease, sublicense, lend, assign, time-share, or transfer, in whole or in part, or provide unlicensed Third Parties access to the Data and Related Materials or portions of the Data and Related Materials, any updates, or Licensee's rights under this Agreement.

Licensee shall not remove or obscure any copyright or trademark notices of ESRI or its Licensors.

Term and Termination: The license granted to Licensee by this Agreement shall commence upon the acceptance of this Agreement and shall continue until such time that Licensee elects in writing to discontinue use of the Data or Related Materials and terminates this Agreement. The Agreement shall automatically terminate without notice if Licensee fails to comply with any provision of this Agreement. Licensee shall then return to ESRI the Data and Related Materials. The parties hereby agree that all provisions that operate to protect the rights of ESRI and its Licensors shall remain in force should breach occur.

Disclaimer of Warranty: The Data and Related Materials contained herein are provided "as-is," without warranty of any kind, either express or implied, including, but not limited to, the implied warranties of merchantability, fitness for a particular purpose, or noninfringement. ESRI does not warrant that the Data and Related Materials will meet Licensee's needs or expectations, that the use of the Data and Related Materials will be uninterrupted, or that all nonconformities, defects, or errors can or will be corrected. ESRI is not inviting reliance on the Data or Related Materials for commercial planning or analysis purposes, and Licensee should always check actual data.

Data Disclaimer: The Data used herein has been derived from actual spatial or tabular information. In some cases, ESRI has manipulated and applied certain assumptions, analyses, and opinions to the Data solely for educational training purposes. Assumptions, analyses, opinions applied, and actual outcomes may vary. Again, ESRI is not inviting reliance on this Data, and the Licensee should always verify actual Data and exercise their own professional judgment when interpreting any outcomes.

Limitation of Liability: ESRI shall not be liable for direct, indirect, special, incidental, or consequential damages related to Licensee's use of the Data and Related Materials, even if ESRI is advised of the possibility of such damage.

No Implied Waivers: No failure or delay by ESRI or its Licensors in enforcing any right or remedy under this Agreement shall be construed as a waiver of any future or other exercise of such right or remedy by ESRI or its Licensors.

Order for Precedence: Any conflict between the terms of this Agreement and any FAR, DFAR, purchase order, or other terms shall be resolved in favor of the terms expressed in this Agreement, subject to the government's minimum rights unless agreed otherwise.

Export Regulation: Licensee acknowledges that this Agreement and the performance thereof are subject to compliance with any and all applicable United States laws, regulations, or orders relating to the export of data thereto. Licensee agrees to comply with all laws, regulations, and orders of the United States in regard to any export of such technical data.

Severability: If any provision(s) of this Agreement shall be held to be invalid, illegal, or unenforceable by a court or other tribunal of competent jurisdiction, the validity, legality, and enforceability of the remaining provisions shall not in any way be affected or impaired thereby.

Governing Law: This Agreement, entered into in the County of San Bernardino, shall be construed and enforced in accordance with and be governed by the laws of the United States of America and the State of California without reference to conflict of laws principles. The parties hereby consent to the personal jurisdiction of the courts of this county and waive their rights to change venue.

Entire Agreement: The parties agree that this Agreement constitutes the sole and entire agreement of the parties as to the matter set forth herein and supersedes any previous agreements, understandings, and arrangements between the parties relating hereto.

Installing the data and software

Getting to Know ArcGIS Desktop includes one CD with exercise data. A trial version of ArcGIS Desktop 10, ArcEditor license (single use) software can be downloaded at www.esri.com/esripress. Use the code printed on the inside back cover of this book to access the download site. **Note:** the code can only be used once.

Installation of the exercise data CD takes about five minutes and requires 153 megabytes of hard disk space.

Installation of the software requires at least 2.4 gigabytes of hard disk space (more if you choose to load the optional extension products). Installation times will vary with your computer's speed and available memory.

If you already have a licensed copy of ArcGIS Desktop 10 installed on your computer (or accessible through a network), do not install the software. Use your licensed software to do the exercises in this book. If you have an older version of ArcGIS installed on your computer, you must uninstall it before you can install the software from the download site.

The exercises in this book have been modified to work with ArcGIS 10. Using another version of the software is not recommended.

Installing the exercise data

Follow the steps below to install the exercise data. Do not copy the files directly from the CD to your hard drive. A direct file copy does not remove write-protection from the files, and this causes data editing exercises not to work. In addition, a direct file copy will not enable the automatic uninstall feature.

1 Put the data CD in your computer's CD drive. A start-up screen will appear. If your auto-run is disabled, navigate to the contents of your CD drive and double-click the Setup.exe file to begin.

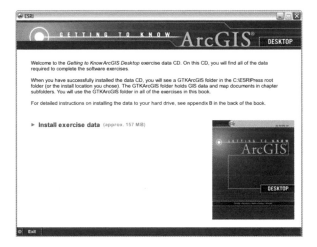

2 Read the welcome screen, then click the Install exercise data link. This launches the Setup wizard.

3 Click Next. Read and accept the license agreement terms, then click next.

4 Accept the default installation folder or click Browse and navigate to the drive or folder location where you want to install the data. If you choose an alternate location, please make note of it as the book's exercises direct you to **C:\ESRIPress\GTKArcGIS**.

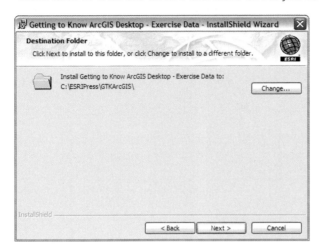

5 Click Next. The installation will take some time. When the installation is finished, you see the following message:

6 Click Finish. The exercise data is installed on your computer in a folder called GTKArcGIS.

If you have a licensed copy of ArcGIS Desktop 10 installed on your computer, you are ready to start *Getting to Know ArcGIS Desktop*. Otherwise, follow the "Installing the software" instructions to install and register the software.

Uninstalling the exercise data

To uninstall the exercise data from your computer, open your operating system's control panel and double-click the Add/Remove Programs icon. In the Add/Remove Programs dialog box, select the following entry and follow the prompts to remove it:

• Getting to Know ArcGIS Desktop Exercise Data

Installing the software

A 180-day trial version of ArcGIS Desktop 10, ArcEditor license (single use) software can be downloaded at www.esri.com/esripress. Use the code printed on the inside back cover of this book to access the download site, and follow the on-screen instructions to download and register the software.

Uninstalling the software

To uninstall the software from your computer, open your operating system's control panel and double-click the Add/Remove Programs icon. In the Add/Remove Programs dialog box, select the following entry and follow the prompts to remove it:

- ArcGIS Desktop

INDEX

Related titles from ESRI Press

The GIS 20: Essential Skills

ISBN: 978-1-58948-256-2

The GIS 20 is a no-nonsense workbook that demonstrates how to perform twenty essential GIS skills as indicated by 500 GIS practitioners. Written for professionals with no time for classroom training, this book can be treated as a weekly self-assignment or an as-needed reference. If you are a GIS beginner, *The GIS 20: Essential Skills* is your best friend.

GIS Tutorial 2: Spatial Analysis Workbook, Second Edition

ISBN: 978-1-58948-258-6

Updated for ArcGIS 10, *GIS Tutorial 2* offers hands-on exercises to help GIS users at the intermediate level continue to build problem-solving and analysis skills. Inspired by the *ESRI Guide to GIS Analysis* book series, this book provides a system for GIS users to develop proficiency in various spatial analysis methods, including location analysis; change over time, location, and value comparisons; geographic distribution; pattern analysis; and cluster identification.

The ESRI Guide to GIS Analysis, Volume 1: Geographic Patterns and Relationships

ISBN: 978-1-87910-206-4

This book presents the necessary tools to conduct real analysis with GIS. Using examples from various industries, this book focuses on six of the most common geographic analysis tasks: mapping where things are, mapping the most and least, mapping density, finding what is inside, finding what is nearby, and mapping what has changed.

Lining Up Data in ArcGIS: A Guide to Map Projections

ISBN: 978-1-58948-249-4

Lining Up Data in ArcGIS is an easy-to-navigate, troubleshooting reference for any GIS user with the common problem of data misalignment. Complete with full-color maps and diagrams, this book presents techniques to identify data projections and create custom projections to align data.

ESRI Press publishes books about the science, application, and technology of GIS. Ask for these titles at your local bookstore or order by calling 1-800-447-9778. You can also read book descriptions, read reviews, and shop online at www.esri.com/esripress. Outside the United States, visit our Web site at www.esri.com/esripressorders for a full list of book distributors and their territories.